T0258154

Advances in Metamaterials

Advances in Metamaterials

Edited by **Austin Doran**

New York

Published by NY Research Press,
23 West, 55th Street, Suite 816,
New York, NY 10019, USA
www.nyresearchpress.com

Advances in Metamaterials
Edited by Austin Doran

International Standard Book Number: 978-1-63238-034-0 (Hardback)

Printed in the United States of America.

Contents

Preface

The world is advancing at a fast pace like never before. Therefore, the need is to keep up with the latest developments. This book was an idea that came to fruition when the specialists in the area realized the need to coordinate together and document essential themes in the subject. That's when I was requested to be the editor. Editing this book has been an honour as it brings together diverse authors researching on different streams of the field. The book collates essential materials contributed by veterans in the area which can be utilized by students and researchers alike.

This book consists of an analysis of the theory, properties and the technological applications of metamaterials for the development of new devices like invisibility cloaks, absorbers and concentrators of EM waves, etc. For developing a new device, it is important to know the electrodynamic features of the metamaterial according to which the device is created. The electromagnetic metamaterials affect EM waves and regulate the surrounding electromagnetic field by changing their permeability characteristics. It is this feature which enables the creation of electromagnetic wave scattering surfaces which utilize metamaterials. This book discusses various aspects related to metamaterials and will be beneficial for both students and experts interested in this field.

Each chapter is a sole-standing publication that reflects each author's interpretation. Thus, the book displays a multi-facetted picture of our current understanding of application, resources and aspects of the field. I would like to thank the contributors of this book and my family for their endless support.

Editor

The Applications of Metamaterials

Antenna Designs with Electromagnetic Band Gap Structures

Dalia M.N. Elsheakh, Hala A. Elsadek and Esmat A. Abdallah
Electronics Research Institute, Giza,
Egypt

1. Introduction

The word "meta", in Greek language, means beyond. It implies that the electromagnetic response of metamaterials (MTMs) is unachievable or unavailable in conventional materials. Many efforts have been done to search for an adequate definition for MTMs. In 2002, J.B. Pendry wrote in a conference paper: "meta-materials, materials whose permeability and permittivity derive from their structure". Later, in 2006, C. Caloz and T. Itoh wrote: "Electromagnetic metamaterials are broadly defined as artificial effectively homogeneous electromagnetic structures with unusual properties not readily available in nature" [1]. Perhaps, a serious obstacle on the road to a universal definition for the term MTMs is the fact that researchers working with these objects do not commonly agree on their most essential characteristics. In [2] and [3], some of the problematic aspects of the non-naturality definition were raised, like the difficulty in separating classical composites from the new class of metamaterials. Another argument against the "not found in nature" property is that it unnecessary excludes impressive examples of natural media that could be called metamaterials par excellence, such as structural colors [4].

MTMs cover an extremely large scientific domain which ranges from optics to nanoscience and from material science to antenna engineering. In this chapter, we focus primarily on the subject of MTMs in the electromagnetic field. Personally, We prefer the definition given by D.R. Smith: Electromagnetic metamaterials are artificially structured materials that are designed to interact with and control electromagnetic waves [5]. The term "artificial" refers to the fact that the electromagnetic response of these materials is dominated by scattering from periodically or amorphously placed inclusions (e.g., metallic or dielectric spheres, wires, and loops) [6].

In the family of MTMs, "left-handed" (LH) media drew an enormous amount of interest. This concept was first put forward by a Russian physicist, Victor Veselago, in 1968, for whom the medium is characterized by a simultaneously negative electric permittivity and negative magnetic permeability [7]. Veselago argued that such media are allowed by Maxwell's equations and that electromagnetic plane waves can propagate inside them, but the phase velocity of such a plane wave is in the opposite direction of the Poynting vector. Hence, some researchers use the term "backward wave media" (BWM) to describe these LH materials [8]. When such media are interfaced with conventional dielectrics, Snell's Law is reversed, leading to the negative refraction of an incident plane wave as shown in figure 1.

Nevertheless, Veselago's conjecture was essentially ignored for thirty years due to the absence of naturally occurring materials or compounds that possess simultaneously negative permittivity and permeability.

In 2000, a metamaterial, based on conducting wires [9] and split-ring resonators (SRRs) [10], was demonstrated to have a negative refractive index over a certain range of microwave frequencies [10-13]. Wires, either continuous or with periodic breaks, can provide a positive or a negative effective permittivity. Planar SRRs or wound coils (also known as Swiss Rolls) can provide a positive or a negative effective permeability. Harnessing the phenomenon of negative refraction, these metamaterials offer a good potential for all kinds of applications, such as "perfect" lens [14], imaging [15], resonators [16], and cloaking [17].

Metamaterials possessing these properties are also frequently named "Negative Refractive Index (NRI)" and "Double Negative (DNG) material". In addition to the materials with simultaneously negative permittivity and negative permeability, the single negative metamaterials have also drawn a great interest. Applications are found for these materials either with a negative permittivity "Epsilon Negative (ENG)" [17] or a negative permeability "Mu Negative (MNG)" [18]. Besides, materials with the properties of "Epsilon near Zero (ENZ)" [19] and "Mu Near Zero (MNZ)", known as "nihility" materials have also been studied. A simple synopsis of these metamaterials can be found in figure 2, where the angular frequencies ωpe and ωpm represent, respectively the electric and magnetic plasma frequency [20]. Up to now, we talked about metamaterials who exhibit their great performances by artificially tailoring the permittivity or permeability. Besides, the term "metamaterial" has also been used by some authors to describe other periodic structures such as electromagnetic bandgap (EBG) structures or photonic crystals, when the period is much smaller in physical size than the wavelength of the impinging electromagnetic wave. The electromagnetic response of such structures is dominated by Bragg-type scattering and involves higher order spatial harmonics (Bloch-Floquet modes) [20]. In this chapter, we focus on such a kind of metamaterial, the so-called "electromagnetic band gap" (EBG). Electromagnetic has received great attention among researchers all over the world because of its immense civilian and defense applications. During the Second World War, the use of radar and thereafter the wide use of microwave communication systems facilitated the transformation from radio to microwave frequency. This dramatic change demanded more advanced materials for high frequency performance and opened up new dimensions in the field of electromagnetic materials. Nano-composites and electromagnetic band-gap structures are examples of metamaterials under right hand rules. Electromagnetic band-gap (EBG) structures have attracted increasing interest in the electromagnetic community. Because of their desirable electromagnetic properties [21], they have been widely studied for potential applications in antenna engineering. Hundreds of EBG papers have been published in various journals and conferences in the last 5 years. EBG are periodic arrangements of dielectric or metallic elements in one, two or three dimensional manners. EBG inhibits the passage of electromagnetic wave at certain angles of incidence at some frequencies. These frequencies are called partial band-gap. At a specific frequency band, EBG does not allow the propagation of wave in all directions and this frequency region is called the complete band-gap or global band-gap [22, 23]. Physicists put the original idea of EBG forward and some recent studies revealed the interesting fact that EBG exists in living organisms. The well known examples are the butterfly wing scales and eyes of some insects. In this case, a metallic like reflection effect is obtained by using refractive index differences.

A multilayer thin film with different refractive indices in animals is a good example for this. Recently, these ideas were undergone a preliminary study for its commercialization such as paints for certain applications.

The concept of electromagnetic band-gap (EBG) structures originates from the solid-state physics and optic domain, where photonic crystals with forbidden band-gap for light emissions were proposed in [27–28] and then widely investigated in the [29–33]. Thus, the terminology, photonic band-gap (PBG) structures, was popularly used in the early days. Since then, a profusion of scientific creativity has been witnessed as new forms of electromagnetic structures are invented for radio frequency and microwaves. EBG can be realized in one, two and three dimensional forms. The dimensionality depends on the periodicity directions. Three dimensional EBG are more appropriate for getting a complete band-gap because they can inhibit waves for all incident angles. The band-gap in EBG is analogous to a forbidden energy gap in electronic crystals. Hence EBG are also termed as photonic crystals (PCs). The first attempts towards three-dimensional structures were realized in the form of face centered cubic (fcc) lattice structures [35]. At the initial stages of EBG research, due to the lack of theoretical predictions, a 'cut and try' approach was adopted in experimentally predicting the band-gap. At the beginning, the investigations of EBG were mainly on wave interactions of these structures at optical frequencies and hence PBG emerged with the name of photonic band-gap structures. Now, vast extensions of EBG at microwave [36], millimetre [37] and sub-millimetre wave frequencies [38] are electromagnetic band-gap (EBG) structures. A periodic structure can give rise to multiple band-gaps. However, it should be noted that the band-gap in EBG is not only due to the periodicity of the structure but also due to the individual resonance of one element. A study revealed the mechanisms to form a band-gap in an EBG [21]. The band-gap formation in EBG is due to the interplay between macroscopic and microscopic resonances of a periodic structure. The periodicity governs the macroscopic resonance or the Bragg resonance. It is also called the lattice resonance. Microscopic resonance is due to the element characteristics and it is called the Mie resonance [20]. When the two resonances coincide, the structure possesses a band-gap having maximum width. Depending on the structural characteristics and polarization of the wave, one resonance mechanism (i.e. either the multiple scattering resonance or the single element scattered resonances) can dominate over the other. The characteristic property of stop bands at certain frequencies enables many applications using EBG. At this stop band, all electromagnetic wave will be reflected back and the structure will act like a mirror. At other frequencies, it will act as transparent medium. This concept is illustrated in figure 3.

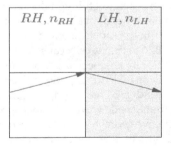

Fig. 1. A negative reflection.

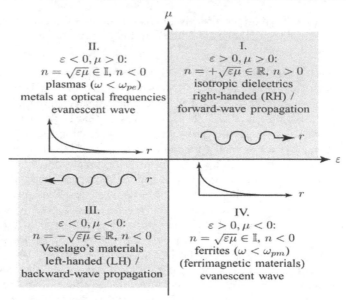

Fig. 2. Permittivity, permeability and refractive index diagram [20].

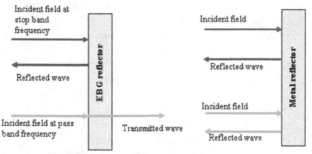

Fig. 3. Diagram illustrating the application of EBG as a mirror and its comparison with a metal reflector [21].

Microstrip antennas mounted on a substrate can radiate only a small amount of its power into free space because of the power leak through the dielectric substrate [32]. In order to increase the efficiency of the antenna, the propagation through the substrate must be prohibited. In this case, the antenna can radiate more towards the main beam direction and hence increase its efficiency. Recently, there has been much interest in the field of Metallo-Dielectric Electromagnetic Band-Gap (MDEBG) structures because of the promising future applications and the important role these artificially engineered periodic materials may play in the field of antennas. The name "Photonic (or Electromagnetic, which is more appropriate for the frequency band of applications) band-gap" has its origin in the fact these structures effectively prevent the propagation of electromagnetic waves within a specific frequency range (the band gap). Two examples of the qualitative geometry of such structures are given in figure 4 [33]. As shown in figure 4, a MDEBG structure is essentially a surface comprising a plurality of elements. Each of the elements is interconnected with each other to form an array of metallic parts embedded in a slab of dielectric. In other words, they are periodical structures of densely packed planar conducting patches separated from a solid metal plane by a dielectric layer. Sometimes metallic pins (or via) are introduced to prevent electromagnetic waves from traveling in the waveguide between the array and the ground. Each unit cell, which is periodically repeated to form the array, essentially behaves as a microwave resonant circuit. The plurality of the resonant elements is parameterized to substantially block surface wave's propagation in the device within a predetermined frequency band gap [35].

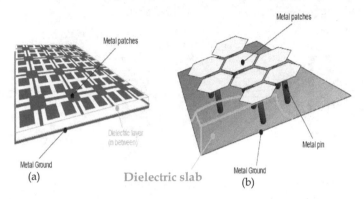

Fig. 4. Simple examples of Metallo-Dielectric EBG structures [32].

The objective of this chapter is to investigate the EM properties of the microstrip antennas based on metallo-dielectric electromagnetic band-gap structures. It is important to emphasize the fact that simple embedded electromagnetic band-gap (EEBG) structures presented in this work are targeting operating frequency band at 0.5 - 20 GHz range. Since simple EEBG structures have impractical geometrical sizes in the 500 MHz to 2 GHz frequency range, more complex EEBG structures need to be employed. In addition, from wireless communication application, miniaturization of electronic systems requires the availability of miniaturized EEBG structures with appropriate patch sizes. The concept introduced in this chapter can be generally applied regardless of the size of the EEBG structures. For wideband radiation, reduction from hundreds of MHz to few GHz either a combination of different methods or use of advanced EEBG structures is the best solution as shown in figure 5 [40]. Figure 5 shows the efficacy range of EEBG structures covered in [40].

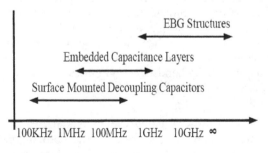

Frequency (MHz)

Fig. 5. Efficacy range of different reduction methods.

2. Disadvantages of metal ground planes and the MDEBG solution

MDEBG structures are useful where the presence of classic electric conductors as antenna ground planes adversely affects the performance of the entire electromagnetic device. As it is known, classic conductive surfaces are extensively used as antenna reflectors: they redirect one half of the radiation into the opposite direction potentially improving the antenna gain by 3 dBi.

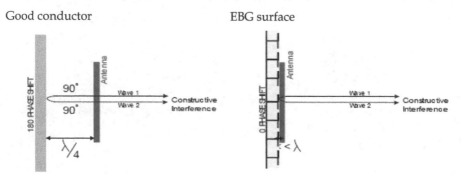

Fig. 6. An antenna separated by ¼ λ from the ground plane (on the left) and the alternative MDEBG layout (on the right).

However, they do have two main disadvantages: first, they reverse the phase of the reflected wave and second, they support propagating surface waves, which can have unwanted effects on the antenna performance. The fact that they reverse the phase 180 degrees is due to the most obvious constraint that the tangent electric field on a classic conductive surface must be zero, so the electromagnetic waves experience a 180 degrees phase shift on reflection. Because of the phase reversal, the image currents cancel the antenna currents, resulting in poor radiation efficiency when the antenna is too close to the conductive surface. This problem is often solved by including a quarter wavelength between the radiating element and the ground plane (see figure 6), but the disadvantage of this solution is the fact that the structure requires a minimum thickness of λ/ 4. It will be shown that, by using the novel MDEBG structures as ground planes, the antenna can be almost be attached to the ground plane, resulting in a useful reduction of volume. It will be proved that, at the frequency where the MDEBG structure does not give any reflection phase shift, a design of MDEBG structures even 8 times thinner than the classic ones (which implement a λ/ 4 spacing between antenna and ground plane) is possible. As stated before, another issue is the propagation of surface waves when normal ground planes are used: these are propagating electromagnetic waves bound to the interface between metal and free space and they will radiate if scattered by bends, discontinuities or surface textures. The unwanted result is a kind of multipath interference, which can be seen as ripples in the radiation pattern (see figure 7) [41].

Again, by using MDEBG structures, it will be shown that surface waves can be suppressed. It follows that, when multiple antennas share the same normal conductive ground plane, like it happens in phased arrays, the above mentioned surface waves may cause undesired mutual coupling between the antennas (see figure 8). Once again, by using MDEBG surfaces structures, it is possible to alter the surfaces properties of the ground plane and avoid this mutual coupling. Therefore, it can be easily understood that the Metallo-Dielectric EBG structure is a useful alternative to antenna classic metallic reflectors.

From above discussion, to realized high performance antenna:

-   A reflector which lacks edge currents that radiate power into the back hemisphere of the antenna is needed;
-   Surface waves on a ground plane associated with an antenna have to be suppressed to provide more efficient antennas, and reduce coupling.

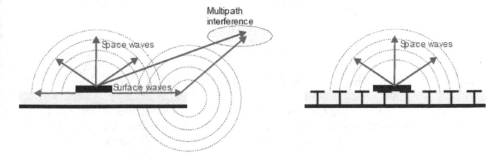

Fig. 7. Multipath interference due to surface waves on a normal ground plane (on the left) and the alternative MDEBG layout (on the right).

Nowadays, one of the most commonly used solution to prevent the propagation of unwanted surface waves is the so called "choke ring" which provides excellent electrical performance for GPS antennas. They are usually very large and heavy. While size and weight are not issues for most base station applications, for a GPS surveyor carrying one around in the field, size and weight are important factors. Moreover, choke rings are also expensive, typically costing thousands of dollars.

3. Metallo EBG structures as novel ground planes for antenna applications

The modern trends in communication systems require wide bandwidth, small size and low profile antennas. A planar microstrip antenna (PMA) is the good candidate for use in UWB wireless technology because they can reach wide impedance bandwidth and nearly omnidirectional azimuthally radiation pattern by different methods. One simple but powerful technique is to replace the cylindrical wires with the plate elements, such as rectangular (square), elliptical (circular), triangular shapes, and others. Another way to increase the impedance bandwidth of the monopole antennas can be achieved by modifying the ground plane. Different shapes of modified ground plane as semi circular is used to increase bandwidth. With the intention to overcome this handicap, a thick, high permittivity substrate is used, and potential surface waves are suppressed applying so called Electromagnetic Band-gap (EBG) [43]. The surface wave propagation is a serious problem of MPA. Surface waves reduce antenna efficiency and gain, limit bandwidth, increase end-fire radiation and cross-polarization levels. To avoid this, the substrate is periodically loaded so that the surface waves cannot propagate along the substrate. Also, other surface wave coupling effects like mutual coupling between array elements and interference with board systems can be suppressed [44].

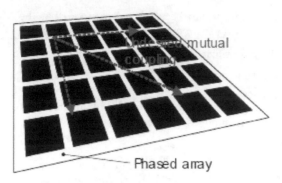

Fig. 8. Mutual coupling for multiple antennas due to the presence of surface waves.

To avoid using costly hybrid technology, innovative design must be developed to successfully integrate antenna with circuitry on high dielectric substrates. The novel RF system architecture reduces size, weight, losses and it is also suitable for integration with micro-electromechanical systems to realize reconfigurable circuits and antennas. Two technologies have been mainly pursued so far to achieve MPA on high-dielectric substrate with optimum performance. One is based on micromachining technology hile the other makes use of the concept of photonic band-gap substrates

A defected ground structure (DGS) has gained significant interests. It rejects certain frequency bands, and hence it is called electromagnetic band-gap (EBG) structures. Liu et al. has presented a novel DGS based meander microstrip line providing a broad stop-band [46]. A novel defected ground structure with islands (DGSI) is proposed by T. Itoh. in [48]. The DGS is realized on the bottom plane with two islands placed at both sides of the microstrip line on the upper plane. Due to their excellent pass and rejection frequency band characteristics, DGS circuits are widely used in various active and passive microwave and millimeter-wave devices such as filters, dividers, couplers, amplifiers, resonators and antennas.

4. Applications

One of the main purposes of this chapter is to deeply go into all the promising possible applications of MDEBG structures by satisfy their challenging design requirements of the antennas. Medical application as image scan is very important nowadays. A very promising way to eradicate the problems created by surface waves in this application (e.g. scan blindness), while at the same time improving performance, is to use electromagnetic band-gap structures instead of standard dielectric antenna substrates. Within the MDEBG, in fact, the unwanted effect of surface waves will be efficiently suppressed.

Another microwave application is high precision GPS. High precision GPS is going to be used in many situations. By very accurately determining the phase of the signal, it is hoped to reach a position accuracy of a few millimeters. However, in order to avoid errors in measurements (due to multi-path), the backward radiated field has to be at least a few orders of magnitude below the frontward field. Such a requirement cannot be obtained if the antenna excites surface waves and electromagnetic band-gap materials could again be used as design solutions. Another application is mobile and wireless communications. As the world goes, wireless, data and voice transmissions are bound to become even more common. A lot of attention is now focused on Bluetooth band hence its implementation in different wireless systems increases in everyday life. Moreover, for other applications like cell-phones, more attention is being paid to the shielding offered by the antenna and the potential health hazard.

Other than that, the EBG structure also can be used as a band reject especially for ultra wide band applications which operate at very wide frequency ranges. The simulation and measurement results for ultra wide band with and without band rejection are investigated in this chapter.

EBG structures are used to prevent some operating modes and make harmonic control. These techniques can increase the usability of antenna systems. The design and simulation of the antennas with EBG structures have recently received more attention. Initial concepts only have been proven with limited fabricated devices. A lot of research efforts is still expected in this field. This chapter is a step to investigate and develop how the EBG structure/ground plane can be used to optimize the antenna performance to satisfy several applications. The investigated antennas are miniaturized with broadband characteristics to be suitable for multi-band/ multi-function operation in wireless communications and medical applications.

5. Problems led to the use of EBG

Antenna designs have experienced enormous advances in the past several decades and they are still under research and development. Many new technologies have emerged in the modern antenna design arena and one exciting breakthrough is the discovery/ development of EBG structures. The applications of EBG structures in antenna designs have become interesting topic for antenna scientists and engineers. The recent explosion in antenna developments has been fueled by the increasing popularity of wireless communication systems and devices. From the traditional radio and TV broadcast systems to the advanced satellite system and wireless local area networks, wireless communications have evolved into an indispensable part of people's daily lives. Antennas play a paramount role in the development of modern wireless communication devices, ranging from cell phones to portable GPS navigators, and from the network cards of laptops to the receivers of satellite TV. A series of design requirements, such as low profile, compact size, broad bandwidth, and multiple functionalities, keep on challenging antenna researchers and propelling the development of new antennas. Progress in computational electromagnetic, as another important driving force, has substantially contributed to the rapid development of novel antenna designs. It has greatly expanded the antenna researchers' capabilities in improving and optimizing their designs efficiently.

Various numerical techniques, such as the method of moments (MOM), finite element method (FEM), and the finite difference time domain (FDTD) method, have been well developed over the years. As a consequence, numerous commercial software packages have emerged. Nowadays with powerful personal computers and advanced numerical techniques or commercial software, antenna researchers are able to exploit complex electromagnetic materials in antenna designs, resulting in many novel and efficient antenna structures with most of required characteristics.

For these reasons, EBG structures and their applications in antennas have become a new research direction in the antenna researches community. It was first proposed to respond to some antenna challenges in wireless communications [40]. For example,

- How to suppress surface waves in the antenna ground plane?
- How to design an efficient low profile antenna near a ground plane?
- How to increase the gain of an antenna?

In the novel RF system architecture of reducing the size, weight, losses and suitability for integration with micro-electro-machine-system lead to achieve reconfigurable antenna with required features. If a conventional substrate is used, then most of the antenna radiation is emitted from the substrate (since it has a higher dielectric constant than air). A big part of this radiation is trapped inside the substrate because of total internal reflection. As a consequence of this, more than 50% of the radiated energy is lost. Also heat dissipation and temperature effects arise in the substrate.

Two technologies have been used to achieve microstrip antenna on high dielectric substrate with optimum performance:

- One is based on micromachining technology.
- Second is photonic band-gap (PBG) substrate.

A photonic crystal essentially behaves much like a band-stop filter, rejecting the propagation of energy over a fixed band of frequencies. An appropriate EBG substrate is selected, then all of the energy can be directed towards the radiating direction (total reflection by the EBG structure), thus improving the antenna directivity and eliminating the substrate heat dissipation [41].

Due to the complexity of the EBG structures, it is usually difficult to characterize them through analytical methods. Instead, full wave simulators that are based on advanced numerical methods have been popularly used in EBG analysis. Dispersion diagram, surface impedance, and reflection phase features are explored for different famous EBG structures. The interaction of antennas and EBG structures are extensively investigated. In summary, the EBG research has flourished since the beginning of this new millennium.

6. Electromagnetic Band-Gap (EBG) structure

6.1 The parameters of EBG (figure 9)

- Permittivity of the dielectric materials used (ε_r)
- Dimensions of the mushroom patches (a)
- Periodicity (P)
- Incident angle of electromagnetic waves (θ_i)

A periodic structure is characterized by the following parameters:

1. $\lambda_r = 2.a$ (1)
2. Shape of individual patches.
3. Filling factor ratio between size of the patches and the periodicity of unit cell (a/P).

For best performance [40].

- For 3D-EBG $0.9 < a/P < 0.95$
- For 2D-EBG $0.65 < a/P < 0.75$

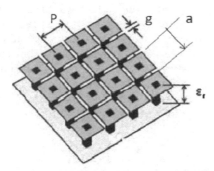

Fig. 9. The parameters of EBG.

6.2 The features of EBG

The main feature of EBG structures is their capability to affect the radiative dynamics within the structure so that there are no electromagnetic modes available within the dielectric. This feature is analogous to periodically arranged atomic lattice of a semiconductor which gives rise to the allowed values of energy that an electron can have at the valence band and at the conduction band, with an energy band-gap separating the two. The optical analogy to this situation is a periodic dielectric structure with alternating high and low values of permittivity, which gives rise to a photonic band-gap [50].

6.3 Applications of EBG

In optical domain new highly efficient opto-electronic devices are considered such as very efficient laser diodes [44, 45], microscale light circuits, multiplexers or demultiplexers based on inhibition of spontaneous emission, photoluminescence, wave-guiding, and superprism phenomenon [46-49]. Studies of frequencies occurring for metal photonic crystals have also shown that the frequency can be controlled and could appear in the microwave region [50]. In the microwave domain, many developments concern the direct control of the electromagnetic energy and its transmission: mirrors, electromagnetic windows, and radiation pattern control. We find also the high impedance material of Sievenpiper et al [51]. They proposed their structure as perfect magnetic wall to reduce the leaky waves in antenna array. The material developed allows the realization of antennas, low loss coplanar lines and compact integrated filter [52-54].

Other applications include duplexers and controllable PBG materials. Due to a certain easiness of fabrication in this frequency domain, the challenge of the electronically controlled photonic crystals has a significant interest. Industrial applications of these crystals are under development, concern mainly aerospace, and telecom domains [43].

6.3.1 Antenna substrates for surface wave suppressions

Surface waves are by-products in many antenna designs. It directs electromagnetic wave propagation along the ground plane instead of radiation into free space, consequently reduce the antenna efficiency and gain. The diffraction of surface waves increases the back lobe radiations, which may deteriorate the signal to noise ratio in wireless communication systems such as GPS receivers. In addition, surface waves raise the mutual coupling levels in array designs as shown in figure 10, resulting in the blind scanning angles in phased array systems. The band-gap feature of EBG structures has found useful applications in suppressing the surface waves in various antenna designs. For example, an EBG structure is used to surround a microstrip antenna to increase the antenna gain and reduce the back lobe as shown in figure 7. In addition, it is used to replace the quarter-wavelength choke rings in GPS antenna designs. Many array antennas also integrate EBG structures to reduce the mutual coupling level more over to increase antenna gain used super-state EBG as shown in figure 10 [40].

Fig. 10. EBG substrate for surface wave suppression with low mutual coupling microstrip array.

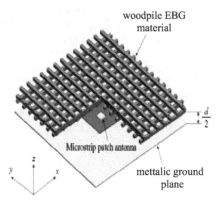

Fig. 11. A high gain resonator antenna design using a woodpile EBG structure.

6.3.2 Antenna substrates for efficient low profile antenna design

Another favorable application of EBG is to design low profile wire antennas with good radiation efficiency, as shown in figure 11, which is desired in modern wireless communication systems. To illustrate the fundamental principle, Table 1 compares the EBG with the traditional PEC ground plane in antenna designs [43].

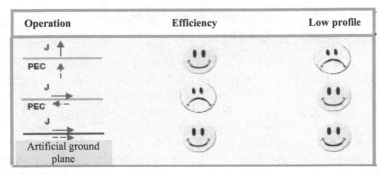

Operation	Efficiency	Low profile
J ↑ ⎯⎯⎯⎯⎯⎯ PEC ↑	☺	☹
J ⎯⎯→ PEC ←‑	☹	☺
J ═══►═ Artificial ground plane	☺	☺

Table 1. Comparison of conventional PEC and artificial ground planes in antenna designs [43].

When an electric current is vertical to a PEC ground plane, the image current has the same direction and reinforces the radiation from the original current. Thus, this antenna has good radiation efficiency, but suffers from relative large antenna height due to the vertical placement of the current. To realize a low profile configuration, one may position an antenna horizontally close to the ground plane. However, the problem is the poor radiation efficiency because the opposite image current cancels the radiation from the original current. In contrast, the EBG surface is capable of providing a constructive image current within a certain frequency band, resulting in good radiation efficiency. In summary, the EBG surface exhibits a great potential for low profile efficient antenna applications as shown in figure 12. Based on this concept, various antennas have been constructed on the EBG ground plane [59–62]. Typical configurations include dipole antenna, monopole antenna, and spiral antenna. EBG surfaces have also been optimized to realize better performance such as multi-band and wideband designs.

Fig. 12. EBG substrate with low profile antenna.

6.4 Advantages of EBG

Utilizing electromagnetic band-gap crystals in a patch antenna with an air gap appears to perform five key functions:

4. Increase operation bandwidth
5. Reduce side-lobe levels
6. Increase front to back (F/B) ratios
7. Increase directivity and consequently gain improvement
8. Harmonic control (suppression of resonance at the harmonic frequencies of the antenna)

7. Classifications of electromagnetic band-gap structures

7.1 Defected Ground Structures (DGS)

A Defected Ground Structure (DGS) is an etched lattice shape, which is located on the ground plane. DGS has arbitrary shapes and is located on the backside metallic ground plane. DGS is realized on the bottom plane with one island placed at both sides of the microstrip line on the upper plane. DGS for the microtrip line, which has etched defects in the backside metallic ground plane, is one hotspot concepts of microwave circuit design nowadays. Compared to photonic band-gap (PBG), DGS has simple structure and potentially great applicability to design microwave circuits such as filters, amplifiers and oscillators. DGSs have gained significant interests. It rejects certain frequency bands, and hence it is called electromagnetic band-gap (EBG) structures as shown in figure 13 [57]. The DGS cell has a simple geometrical shape, such as rectangle. Novel fractal DGS cell is proposed. Its band-gap and slow-wave characteristics are better than the conventional ground plane. DGSs have gained quite significance in filter design [58] showing optimal pass-band and stop-band responses plus sharp selectivity and ripple rejection. Application of CPW-based spiral-shaped DGS to MMIC for reduced phase noise oscillator [57], active devices (BJT and FET) can also be mounted using DGS technique. High amount of isolation is achieved in microstrip diplexer and harmonic control can also be achieved on microstrip antenna structures using DGS. Figure 14 gives the schematic of such a DGS with its approximate surface area. A novel DGS based meander microstrip line providing a broad stop band is presented in [59]. Novel Defected Ground Structures with Islands (DGSI) is proposed in [58]. Careful selection of the line width guarantees 50Ω characteristic impedance (Z_0).

The EM simulation results of the DGSI are compared with circuit simulation results using extracted parameters; showing excellent agreement between the two in wide band. Examination of stop band characteristics is studied using concentric circular rings in different configurations. Metallic backing significantly reduces interference effects, harmonics and phase noise. Several novel 1D DGS are presented for microwave integrated circuits (MIC), monolithic MIC (MMIC), low temperature cored ceramic (LTCC) including RF front-end applications. Significant change in the characteristics including slow-wave factor (SWF) of periodic structures like transmission lines is achieved using quite a few unconventional DGS like spiral-shaped and vertically periodic DGS. Vertically periodic DGSs (VPDGS) have been used in reducing the size of MIC and amplifiers, thus increasing SWF significantly. Harmonic control can also be achieved on microstrip antenna structures using 1-D DGS [59].

Table 2 presents the difference between defected ground plane structure and band-gap structure. The characteristics of the defected ground structure are:

-   Disturbs shielding fields on the ground plane.
-   Increases effective permittivity.
-   Increases effective capacitance and inductance of transmission line.
-   Has one-pole LPF characteristics (3dB cutoff and resonance frequency).
-   Size reduction for the component.

Comparisons	Electromagnetic Band-gap Structure (EBG)	Defected Ground Structure DGS
Geometry	Periodic etched structure	One or few etched structure
Microwave Circuit properties	Similar	Similar
Equivalent Circuit extraction	Very difficult	Relatively simple

Table 2. Comparison between defected ground structure and EBG..

7.2 Photonic Band-Gap (PBG) structures

7.2.1 Historical background of PBG

Photonic band-gap (PBG) structures are periodic structures that manipulate electromagnetic radiation in a manner similar to semiconductor devices manipulating electrons. Semiconductor material exhibits an electronic band-gap where there are electrons cannot exist. Similarly, a photonic crystal that contains a photonic band-gap does not allow the propagation of electromagnetic radiation with specific frequencies within the band-gap [60]. This phenomenon results from the destructive Bragg diffraction interference due to the periodic boundary conditions of PBG structures. This property has a significant importance in many microwave and optical applications to improve their efficiency. The photonic band-gap structures were first investigated in [61] by Yablonovitch. He introduced vector-spherical-wave expansion method or the vector Koringa-Kohn-Rostker (KKR) method to calculate the dispersion relation and the transmittance for the regular array of dielectric spheres. This was the first self-consistent treatment of the electromagnetic eigen modes in 3-D dielectric systems with large periodic modulation of the dielectric constant. A remarkable step was made by Yablonovitch who pointed out the possibility of the realization of photonic band-gap, localized defect modes, and their applications to various optoelectronic devices. His ideas stimulated many researchers, and energetic research activities including his own studies were initiated. The process to realize the photonic band-gap is described in [62]. Lee et al [49] discussed the strong localization of electromagnetic waves in disordered PBG structures.

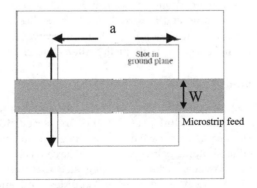

Fig. 13. Schematic diagram of a unit DGS cell.

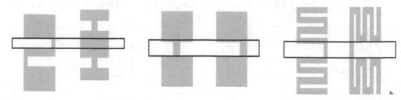

Fig. 14. Different shapes of DGS structures.

Since then, many researchers in various fields such as physics, electronics, waves, optics, fabrication, and chemistry have been engaged in the realization of photonic band-gap, localized defect modes, and other microwave and optical properties peculiar to the PBG structures [63]. They have also been collaborating to make new structures and measure their properties. PBG structures were initially applied to optical applications like high-quality optical mirrors and resonators. One of the most exciting projected applications of PBG structures is the production of optical circuits by manipulating light with optical waveguides. Currently, fiber-optic signals must be converted to electronic signals in order to be analyzed by computers or other devices connected to the optical line. This slows the signal considerably and the electronic circuitry is fairly inefficient as well. A device that runs completely on light would revolutionize technology in several areas, leading eventually to an all optical network (AON).

The integration of photonic components such as lasers, detectors, couplers, and waveguides is still at a very primitive stage due to difficulties with implementing integrated optical components smaller than certain sizes. For instance, small bends or curves of waveguides lead to the leakages of optical signals, so the bends have to be bigger than certain length. PBG structures are proposed as a possible solution to this problem. Adding a defect in a PBG structure opens a path along which electro-magnetic radiation can propagate. Because the PBG structure can be tailored to completely reflect certain frequencies, it is possible to turn corners with light at a distance of the light's wavelength order. Another application of PBG structures is the use of these crystals to greatly improve the efficiency of lasers and light-emitting diodes. Due to the scalability of PBG structures, several applications at microwave and millimeter-wave frequencies have been developed [64]. Other applications such as resonant cavities can be designed by using PBG structures are also presented. Applications based on the guiding and localized mode properties of the PBG structures are given. Some of these common applications in both microwaves and optics are power splitters, switches, directional couplers, high quality filters, and channel drop filters.

7.2.2 Basic PBG structures

Figure 15, 16 shows different configurations of PBG structures composed of two different materials. These configurations include 1D, 2D and 3D periodicities. The two different materials of a PBG structure can be two different dielectric materials or a metal and a dielectric material. Metallic photonic band-gap systems have received far less attention than dielectric PBG structures. However, it has been suggested that periodic metallic structures have important applications, such as cavities, waveguides, and antennas [65]. The main advantage of dielectric PBG structures is that they do not include metallic loss which is usually the dominant loss at high frequencies. Dielectric materials that can be used to construct PBG structures are widely available for almost all frequency bands from ultraviolet to microwave range which facilitate the design and fabrication. Many basic characteristics are common in both ordinary and photonic crystals. These characteristics are utilized to build the fundamental theories of photonic crystals. However, a major difference between ordinary and photonic crystals is the scale of the lattice constant. In the case of ordinary crystals, the lattice constant is in the order of angstroms. On the other hand, lattice constant is constructed with millimeter dimensions for microwave range. Figure 16 shows examples of different PBG structures used in the design of different applications. Various technologies have been developed and applied for manufacturing PBG structures in the last ten years such as stacking slabs of 2D materials for 3D structures implementations. For practical implementation of an infinite 2D PBG structure, a 2D PBG slab of finite thickness is surrounded by two perfect electric conductor (PEC) plates. This modification is implemented to guide the waves in the normal direction to the periodic cell. In optical applications, confinement of electromagnetic waves inside a 2D PBG slab can be obtained by surrounding the slab by a different material of another dielectric constant [66].

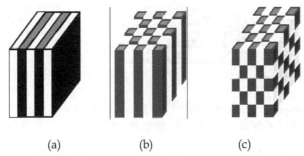

(a)   (b)   (c)

Fig. 15. Examples of different PBG, (a) 1D, (b) 2D and (c) 3D configurations.

Fig. 16. Typical examples of three-dimensional photonic crystals.

i.   The Unit Cell

As stated in the American Heritage Dictionary of the English Language, definition of a crystal (unit cell) is "a homogenous solid formed by a repeating, three-dimensional pattern of atoms, molecules, or shapes, having fixed distances between constituent parts". This replicating pattern is referred to as the unit cell of a crystal.

Figure 17 illustrates the unit cell for the circular crystal lattice. The unit cell contains all the pertinent information of the crystal such as the crystal geometry (shape, thickness, etc.), material properties (dielectric or magnetic), and the lattice spacing (shown as the dimension "p". It is this replicating unit cell that provides the periodicity in the crystal, and controls the location and extent of the band-gap.

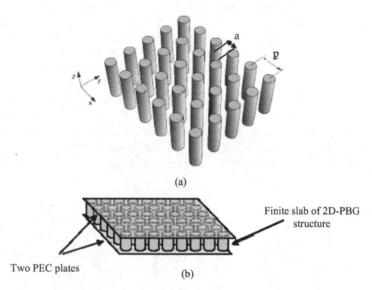

(a)

Finite slab of 2D-PBG structure

Two PEC plates

(b)

Fig. 17. Practical implementation of 2D PBG structure, (a) Finite slab of 2D PBG structure surrounded by air, (b) Finite slab of 2D PBG structure surrounded by two PEC plates.

ii.   The Band-gap

As a first order approximation, a band-gap is obtainable in a high dielectric material with integrated photonic crystals when an incident electromagnetic field propagates with a guide-wavelength, approximately equal to the lattice spacing of the crystal.

$$\lambda_r = \frac{c}{f_r \sqrt{\varepsilon_r}} = 2a \tag{2}$$

This rough approximation locates the center of the band-gap, which can extend higher than ±10% of the center frequency for high dielectric constant materials [67] by using a typical transmission coefficient (S$_{21}$) plot for a 2-port network, such as a microstrip transmission line. This curve, figure19, illustrates that as port 1 of the transmission line is excited, over 15dB of attenuation is experienced as the energy propagates from port 1 to port 2. Thus, the photonic crystal introduces a stop-band filter response. Indeed, the formation of the band-gap is heavily dependent on (1) the periodicity of the crystal, (2) the refractive index (dielectric constant) ratios between the base material (the substrate as a whole) and the impurities that form the crystal. Typically, the refractive index ratio must be at least 2:1 (substrate-to-impurity) ratio for the band-gap to exist. For the 2-D triangular structure, the broadest band-gap is obtainable when the impurities (the cylindrical post) are of air (ε_r =1), while the base material is a high dielectric constant (for example, ε_r =10). A 10:1 dielectric ($\sqrt{10}$ = 3.16:1 refractive index) ratio would satisfy the index requirement and form a broad band-gap, with proper crystal spacing. This explains the need for a high dielectric substrate for a patch antenna designed on a photonic crystal substrate [68].

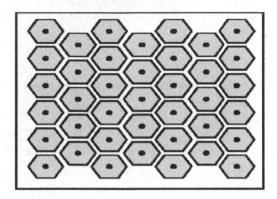

Fig. 18. Top view of the high-impedance surface.

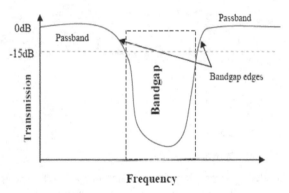

Fig. 19. Transmission loss plot illustrates band gap in a microstrip transmission line at microwave frequencies.

iii. Defects in Periodicity

A photonic crystal essentially behaves much like a band-stop filter, rejecting the propagation of energy over a fixed band of frequencies. However, once a defect is introduced such that it disrupts the periodicity in the crystal, an area to localize or "trap" electromagnetic energy is established. In this region, a pass-band response is created. This ability to confine and guide electromagnetic energy has several practical applications at microwave frequencies as filters, couplers, and especially antennas. This rather simple concept of placing defects in a photonic crystal structure introduces a new methodology in the design of microstrip (patch) antennas. The idea is to design a patch antenna on a 2D photonic crystal substrate, where the patch becomes the "defect" in the crystal structure.

In this case, crystal arrays of cylindrical air holes are patterned into the dielectric substrate of the patch antenna. By not patterning the area under the patch, a defect is established in the photonic crystal, localizing the EM fields. Surface waves along the XY plane of the patch are forbidden from forming due to the periodicity of the photonic crystal in that plane. This prevention of surface waves improves operational bandwidth and directivity, while reducing side-lobes and coupling, which are common concerns in microstrip antenna designs [68]. Using these concepts, a photonic crystal patch antenna was developed.

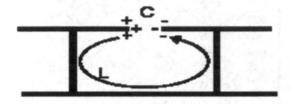

Fig. 20. Origin of the equivalent circuit elements.

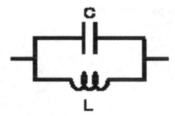

Fig. 21. Equivalent circuit model for the high-impedance surface

7.3 High Impedance Electromagnetic Surface (HIES)

7.3.1 Introduction and background of HIES

A new type of metallic electromagnetic structure has been developed. It is characterized by high surface impedance. Although it is made of continuous metal and conducting DC currents, it does not conduct AC currents within a forbidden frequency band. Unlike normal conductors, this new surface does not support propagating surface waves, and it reflects electromagnetic waves with no phase reversal. The geometry consists of a metal sheet, textured with a two-dimensional lattice of resonant elements, which act as a two-dimensional filter to prevent the propagation of electric currents. The surface can be described using a lumped parameter circuit model, which accurately predicts many of its electromagnetic properties. This unique material is applicable to a variety of electromagnetic problems, including new kinds of low-profile antennas. By incorporating a special texture on a conducting surface, it is possible to alter its radio-frequency electromagnetic properties [69].

In the limit where the period of the surface texture is much smaller than the wavelength, the structure can be described using an effective medium model, and its qualities can be summarized into a single parameter: the surface impedance. A high-impedance surface, shown in figure 18, consists of an array of metal, protrusions on a flat metal sheet. They are arranged in a two-dimensional lattice, and are usually formed as metal plates, connected to the continuous lower conductor by vertical posts as shown in figure 20. They can be visualized as mushrooms or thumbtacks or other shapes protruding from the surface. High Impedance Surfaces as two dimensional EBG structures can be used as microstrip antenna substrate to eliminate the surface wave [70].

7.3.2 High-impedance surfaces properties

The properties of the new high-impedance surface are similar to those of the corrugated slab. The quarter-wavelength slots have simply been folded up into lumped elements, capacitors and inductors that are distributed in two dimensions. The two-dimensional array of resonant elements can be explained using a simple circuit model. The capacitance is due to the proximity of the top metal patches, while the inductance originates from current loops within the structure, as shown in figure 20. The electromagnetic properties of the surface can be predicted by using an equivalent LC circuit, shown in figure 21. The impedance of a parallel resonant LC circuit, given in Eq. 3, is qualitatively similar to the tangent function that describes the impedance of the corrugated surface.

$$Z_s = \frac{j\omega L}{1 - \omega^2 LC} \tag{3}$$

It is inductive at low frequencies, and thus supports TM surface waves. It is capacitive at high frequencies, and supports TE surface waves. In a narrow band around the LC resonance, the impedance is very high. In this frequency range, currents on the surface radiate very efficiently, and the structure suppresses the propagation of both types of surface waves. Having high surface impedance, it also reflects external electromagnetic waves without the phase reversal that occurs on a flat conductor. By using lumped elements, we retain the reflection phase and surface wave properties of the quarter-wave corrugated slab, while reducing the overall thickness to a small fraction of a wavelength [72-74].

7.3.3 Improved surface wave current

Surface waves are excited on microstrip antenna when the substrate $\epsilon_r > 1$. Besides end fire radiation, surface waves give rise to coupling between various elements of an array. Surface waves are launched into the substrate at an elevation angle θ lying between $\pi / 2$ and $\sin^{-1}(1 / \sqrt{\varepsilon_r})$. These waves are incident on the ground plane at this angle, get the reflected from there, then meet the dielectric-air interface, which also reflect them. Following this zig-zag path, they finally reach the boundaries of the microstrip structure where they are reflected back and diffracted by the edges giving rise to end-fire radiation [94]. On other way in the boundary, if there is any other antenna in proximity, the surface wave can become coupled into it. Surface waves will decay as $1 / \sqrt{\varepsilon_r}$ so that coupling also decreases away from the point of excitation. Surface wave are TM and TE modes of the substrate. These modes are characterized by waves attenuating in the transverse direction (normal to the antenna plane) and having a real propagation constant above the cut-off frequency. The phase velocity of the surface waves is strongly dependent on the substrate parameters h and ϵ_r. Figure 22 shows the propagation of the surface wave in microstrip antenna [70].

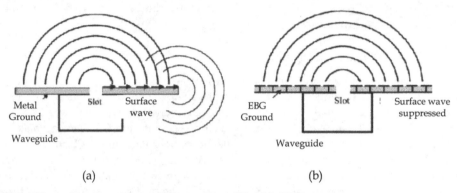

(a)   (b)

Fig. 22. (a) The substrate without EBG structure, (b) with EBG structure.

Surface wave propagation is a serious problem in microstrip antennas. It reduces antenna efficiency and gain, limits bandwidth, increases end-fire radiation, increases cross-polarization levels, and limits the applicable frequency range of microstrip antennas. Two solutions to the surface wave problem are available now. One of the approaches is based on the micromachining technology in which part of the substrate beneath the radiating element is removed to realize a low dielectric constant environment for the antenna. In this case the power loss through surface wave excitation is reduced and coupling of power to the space wave is enhanced. The second technique relies on electromagnetic band-gap structure (EBG) engineering. In this case, the substrate is periodically loaded so that the surface wave dispersion diagram presents a forbidden frequency range (stop-band or band-gap) about the antenna operating frequency. Because the surface waves cannot propagate along the substrate, an increase amount of radiating power couples to the space waves. Also, other surface wave coupling effects like mutual coupling between array elements and interference with onboard systems are now absent [43].

7.3.4 Artificial Magnetic Conductors (AMC)

Background of AMC

Artificial magnetic conductors (AMC), also known as high-impedance surfaces [76] as shown in figure 23(a) and (b), have received considerable attention in recent years. An AMC is a type of electromagnetic band-gap (EBG) material or artificially engineered material with a magnetic conductor surface for a specified frequency band. AMC structures are typically realized based on periodic dielectric substrates and various metallization patterns. Several types of AMC ground planes have already been extensively studied. AMC surfaces have two important and interesting properties that do not occur in nature and have led to a wide range of microwave circuit applications.

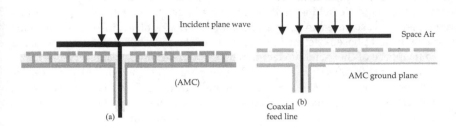

Fig. 23. Horizontal patch antenna on Artificial Magnetic Conductors ground plane.

In the artificial surfaces such as artificial magnetic conductors, the high-impedance surface has proven useful as an antenna ground plane. AMC takes advantage of both the suppression of surface waves and the unusual reflection phase. As a result of the suppression of surface waves, an antenna on Artificial Magnetic Conductors produces a smoother radiation profile than a similar antenna on a conventional metal ground plane, with less power wasted in the backward direction. This can be applied to a variety of antenna designs, including patch antennas, which often suffer from the effects of surface waves. For phased arrays, the suppression of surface waves can reduce inter-element coupling, and help to eliminate blind angles. AMC is particularly applicable to the field of portable hand-held communications, in which the interaction between the antenna and the user can have a significant impact on antenna performance. Using this new ground plane shown in figure 23 as a shield between the antenna and the user in portable communications equipment can lead to higher antenna efficiency, longer battery life, and lower weight.

Parameter	Dimension Change	Center Frequency	Band edge Effect	Bandwidth
Effect of h_1	h_1	Negative Correlation	Lower Side Band	Positive Correlation
Effect of ε_r	ε_r	Negative Correlation	Upper Side Band	Negative Correlation
Effect of radius	r_s	Negative Correlation	Upper Side Band	Negative Correlation
Effect of gap	g	Positive Correlation	Upper Side Band	Positive Correlation
Via position		Negative correlation	Upper side band	Negative Correlation
Via type		No Effect	No Effect	No Effect

Table 3. Parameters Analysis of EBG Design.

AMC surfaces have very high surface impedance within a specific limited frequency range, where the tangential magnetic field is small, even with a large electric field along the surface [73]. Therefore, an AMC surface can have a reflection coefficient of +1 (in-phase reflection). Generally, the reflection phase is defined as the phase of the reflected electric field which is normalized to the phase of the incident electric field at the reflecting surface. It can be called in-phase (or out-of-phase) reflection, if the reflection phase is 0 (or not). In practice, the reflection phase of an AMC surface varies continuously from +180 to -180 relative to the frequency, and crosses zero at just one frequency (for one resonant mode). The useful bandwidth of an AMC is generally defined as 135 to 45 on either side of the central frequency. Thus, due to this unusual boundary condition, in contrast to the case of a conventional metal plane, an AMC surface can function as a new type of ground plane for low-profile wire antennas, which is desirable in many wireless communication systems.

Difference	EEBG	UP-EBG
Substrate structure		
Impedance	Low impedance Surface (LIS)	High impedance Surface (HIS)
Defacement shapes		
Equivalent circuit		
Resonance frequency	$f_0 = \dfrac{1}{2\pi\sqrt{L(C_1 + C_2)}}$	$f_0 = \dfrac{1}{2\pi\sqrt{LC_1}}$
Bandwidth	$\pm\Delta\omega = \omega_o C_1 \dfrac{(1 - L_1 L_2 \omega^2_o)Z_0}{4C}$	$\Delta\omega / \omega_o = \dfrac{Z_o}{\eta}$

Table 4. Comparison between different characteristics of electrical equivalent circuit for both embedded EBG and UP- EBG.

8. Design of single patch antennas with DGS

8.1 Miniaturized design consideration

A novel design method using the first iteration of fractal carpet gasket is utilized as DGS cell to study the size miniaturization. Figure 24(a) shows schematics of the proposed DGS cell. The DGS is etched on the back side of the metallic ground plane as shown in figure 24(b). The substrate is RT/Duroid with 0.813mm thickness and dielectric constant ε_r=3.38. The feed line width w =1.85mm is chosen for the characteristics impendence of 50 Ω microstrip line at frequency 5.25GHz. The defect unit is with length L=12mm and width W=9mm, L_g=1mm, L_s=1mm, W_g=1mm and W_s=3mm. The effect of one as well as two unit cells of rectangular DGS (RDGS) on inset feed MPA performance is studied by etching the defect from certain reference point as a start at distance X=0 mm as shown in figure 24(b) and gradually increase X to change the position of DGS with step 2mm until DGS is far away from the image projection under the radiating surface. For further reduction two unit cells of the proposed RDGS are used, which are etched in face to face with separation 3mm. it is noted that the maximum size reduction is 35% in case of one unit cell at defected displacement X=10mm from antenna edge. In case of two unit cells the maximum reduction ratio reached 53% at same defect position [82].

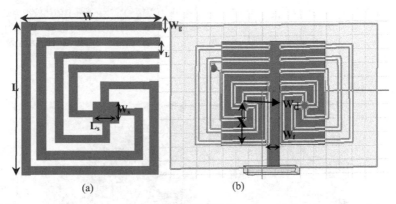

(a)   (b)

Fig. 24. (a) The geometry of one unit cell of DGS, (b) the proposed antenna prototype

8.2 Multiband design consideration

In many applications it is also important to design single feed antenna for multiband resonant frequencies. Spiral defected ground structure DGS is used to provide both multi resonant frequencies and compact size. Figure 24(a) shows schematic of the proposed one cell spiral DGS. Spiral structures, however, are known to produce large cross polarization. Therefore to completely eliminate the cross polarization, a four-arm spiral is explored, as shown in figure 24(b). Four spiral branches, each with a 0.01 $\lambda_{5.2GHz}$ width, split from the center and rotate outwards are used in the design.

As it may be seen, this geometry is symmetric not only in +/-x or +/-y directions, but also in x and y directions. A design using one unit cell of the spiral DGS is simulated and it is clear that the effect of increasing distance X on antenna performance of one cell as well as two unit cells. As distance X increases the resonant frequencies are reduced until certain distance and after that the trend is reversed. Maximum reduction in the fundamental resonant frequency was achieved at distance X=9mm (0.16$\lambda_{5.2GHz}$) and DGS shifted from width center W_d by 3mm (0.083$\lambda_{5.2GHz}$). More reduction in resonant frequency was achieved by using two unit cells of spiral DGS which are placed face to face and at a separation distance from center to center $2W_d$ equal to 10mm (0.18λ_0) with separation W_g equal to 3mm (0.054$\lambda_{5.2GHz}$). We use same previous substrate and antenna dimensions are 15 X16 mm^2 and DGS is etched in the bottom of the metallic ground plane as in figure 24 (b). The dimension of the largest arms length L=13.25mm and largest width W=11.25mm, air-gap g=0.5mm and L_g= 2mm with spacing patch L_s=W_s=1mm with 0.5mm inner square is used in the design. The line width W_f=1.85mm is chosen for the characteristics impendence of 50 Ω microstrip line at frequency 5.25GHz. The fundamental MPA resonance frequency is reduced by using one unit cell spiral DGS which is shifted from length center by 0.011λ_0. In case of using two unit cells, maximum reduction in resonant frequency occurs when the cells are face to face and centered under radiating plate but far from width center by 0.05λ_0.

8.3 Simulation and measurement of miniaturized multi-band antenna

The final two proposed antenna configurations were fabricated as shown in figure 25 by using photolithographic techniques. The reduction in antenna resonant frequency is due to increase in both electrical and magnetic coupling from ground to radiating antenna plate. The measured and simulated results are shown in figure 26. Second proposed antenna was fabricated also and the multi-band operation and reducing antenna size was achieved and the comparison between simulation and measurement of reflection coefficient is shown in figure 27 [81].

Fig. 25. Photo of the fabricated two antennas.

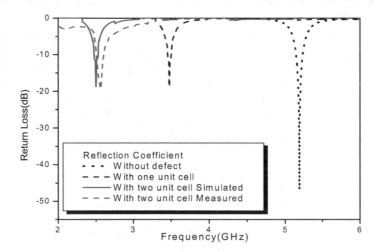

Fig. 26. Comparison between measured and simulated reflection coefficient of antenna without and with one and two gasket unit cell

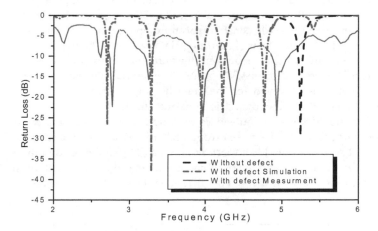

Fig. 27. Comparison between measured and simulated reflection coefficient of antenna without and with two unit cell of spiral DGS simulated and measured.

9. Design of single patch antennas with EBG

The conventional half-wavelength size is relatively large in modern portable communication devices. Various approaches have been proposed, such as using shorting pins, cutting slots, and designing meandering microstrip lines. Increasing the dielectric constant of the substrate is also a simple and effective way in reducing the antenna size. Applications of MPAs on high dielectric constant substrate are of growing interest due to their compact size and conformability with monolithic microwave integrated circuits (MMIC). However, there are several drawbacks with the use of high dielectric constant substrate, namely, narrow bandwidth, low radiation efficiency, and poor radiation patterns, which result from strong surface waves excited in the substrate. The narrow bandwidth can be expanded by increasing the substrate thickness, which, however, will launch stronger surface waves. As a result, the radiation efficiency and patterns of the antenna are further degraded. To quantify this phenomenon, a comparative study of MPAs on substrates with different dielectric constants and different thicknesses is performed in this section. Table 5 illustrates the four samples under study. Two of them with low dielectric constant substrate (ε_r=2.2) and the other two are built on the high dielectric constant substrate (ε_r=10.2).

Example	Patch size mm^2	Dilectric constant (ε_r)	Height (mm)
1	18 X 10	2.2	1
2	16X 13	2.2	2
3	9 X 6	10.2	1
4	8 X 6	10.2	2

Table 5. The antennas parameters.

Figure 28 shows the simulated S_{11} of these four structures. By tuning the patch size and the feeding probe location, all the antennas match well to 50 Ω around 5.1 GHz. It is noticed that the patch sizes on high dielectric constant substrate are remarkably smaller than those on low dielectric constant substrate as shown in table 5, which is the main advantage of using high dielectric constant substrate. However, the antenna bandwidth (S_{11} < 10 dB) on 1 mm substrate height is decreased from 1.38% to 0.61% when the ε_r is increased from 2.2 to 10.2. Similar phenomenon is observed for the 2mm height, the bandwidth is decreased from 2.40% to 1.71%. For the same dielectric constant substrates, the antenna bandwidth is enhanced when the thickness is doubled. For example, the antenna bandwidth on the high dielectric constant substrate is increased from 0.61% to 1.71% when the substrate thickness is increased from 1 mm to 2 mm. It's important to point out that the bandwidth of example (4) is even larger than that of example (1), which means that the bandwidth of MPAs on high permittivity substrate can be recovered by increasing the substrate thickness.

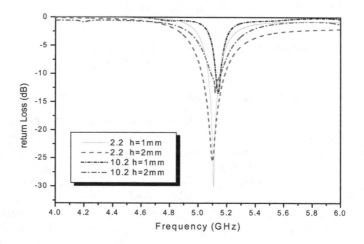

Fig. 28. Return loss comparison of patch antennas with different dielectric constants and substrate height.

Figure 29 compares the H-plane radiation patterns of these four antennas. A finite ground plane of $\lambda \times \lambda$ size is used in the simulations, where λ is the free space wavelength at 5.1 GHz. The antennas on the high dielectric constant substrates exhibit lower directivities and higher back radiation lobes than those on the low dielectric substrates. For antennas on the same dielectric constant substrate, when the thickness increases, the antenna directivity decreases, especially for those on high dielectric constant substrates. Similar observations are also found in the E-plane patterns.

These phenomena can be explained from the excitation of surface waves in the substrate. When a high dielectric constant and thick substrate is used, strong surface waves are excited. This causes reduction of the radiation efficiency and directivity. In addition, when the surface waves diffract at the edges of the ground plane, the back radiation is typically increases.

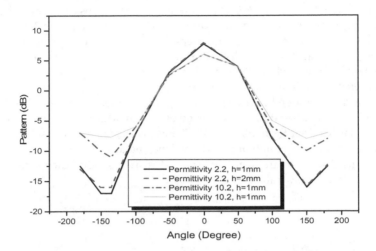

Fig. 29. H-Plane radiation pattern of patch antennas with different dielectric constants and substrate heights.

i.   Gain Enhancement of a Single Patch Antenna

To overcome the drawbacks of using the thick and high dielectric constant substrate, several methods have been proposed to manipulate the antenna substrate. One approach suggested is to lower the effective dielectric constant of the substrate under the patch using micromachining techniques [94]. A shortcoming of this approach is the larger patch size than that on an unperturbed substrate. Another approach is to surround the patch with a complete band gap structure or synthesized low dielectric constant substrate so that the surface wave's impact can be reduced. A MPA design is proposed that does not excite surface waves. In this section, EBG structure is applied in patch antenna design to overcome the undesirable features of the high dielectric constant substrates while maintaining the desirable features of utilizing small antenna size.

ii.   Patch Antenna Surrounded by EBG Structures

Figure 30 sketches the geometry of a MPA surrounded by a mushroom-like electromagnetic band gap (EBG) structure. The EBG is designed so that its surface wave band gap covers the antenna resonant frequency. As a result, the surface waves excited by the patch antenna are inhibited from propagation by the EBG structure. To effectively suppress the surface waves, four rows of EBG cells are used in the design. It is worthwhile to point out that the EBG cell is very compact because of the high dielectric constant and the thick substrate employed. Therefore, the ground plane size can remain small, such as $1\lambda \times 1\lambda$. For comparison, MPA designed on a step-like substrate is investigated, as shown in figure 31. The idea is to use a thick substrate under the patch to keep the antenna bandwidth and use a thin substrate around the patch which reduces the surface waves. The distance between the patch and the step needs to be carefully chosen. If the distance is too small, the resonant feature of the patch will change and the bandwidth will decrease. However, when the distance is too large, it cannot reduce the surface waves effectively. To validate the above design concepts, four antennas were simulated on RT/Duroid 6010 (ε_r = 10.2) substrate with a finite ground plane of 52 × 52 mm². Two of them are normal patch antennas built on 1.27 mm and 2.54 mm thick substrates as references. The step-like structure stacks two 1.27 mm thick substrates under the patch and the distance from the patch edge to the step is 10 mm. The EBG structure is built on 2.54 mm height substrate and the EBG patch size is 2.5 × 2.5 mm² with 0.5 mm separation. Figure 32 compares the measured S_{11} results of these four antennas. All the four patches are tuned to resonate at the same frequency 5.1 GHz. It is noticed that the patch on the thin substrate has the narrowest bandwidth of only 1% while the other three have similar bandwidths of about 3–4%. Thus, the thickness of the substrate under the patch is the main factor determining the impedance bandwidth of the antenna. The step substrate and the EBG structure, which are located away from the patch antenna, have less effect on the antenna bandwidth.

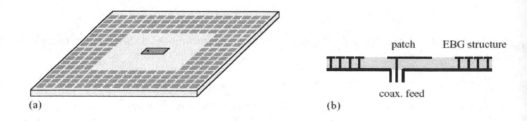

(a)   (b)

Fig. 30. Patch antenna surrounded by a mushroom-like EBG structure: (a) geometry and (b) cross section.

The antenna on the height 2.5mm has the lowest front radiation while its back radiation is the largest. When the substrate thickness is reduced, the surface waves become weaker and the radiation pattern improves. The step-like structure exhibits similar radiation performance as the antenna on the thin substrate. The best radiation performance is achieved by the EBG antenna structure. Due to successful suppression of surface waves, its front radiation is the highest, which is about 3.2 dB higher than the thick case. Since the surface wave diffraction at the edges of the ground plane is suppressed, the EBG antenna has a very low back lobe, which is more than 10 dB lower than other cases. Table 6 lists the simulated results of these antennas. Note that the radiation patterns are normalized to the maximum value of the EBG antenna.

Antenna	Bandwidth %	Front Radiation (dB)	Back Radiation (dB)
Thin	1	-2.3	-15.5
Thick	4	-3.2	-12
Step stair	4.7	-2	-14
EBG	3	0	-25

Table 6. Simulated performance of four different MPA designs on the high dielectric constant substrate.

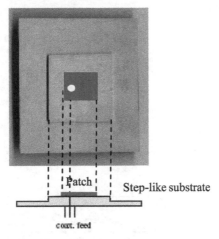

Fig. 31. Patch antenna on a step-like substrate.

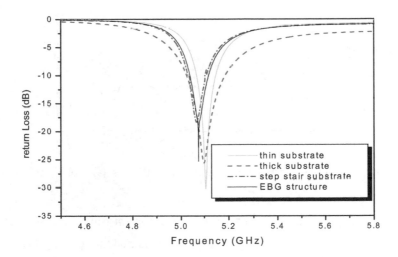

Fig. 32. Comparison of the measured return loss of the four MPA structures.

It is also interesting to notice that in the E-plane the beam-width of the EBG case is much narrower than the other three cases whereas in the H-plane it is similar to other designs. The reason is that the surface waves are mainly propagating along the E-plane as shown in figure 33. Once the EBG structure stops the surface wave propagation, the beam becomes much narrower in the E-plane. From above comparisons it is clear that the EBG structure improves the radiation performances of the patch antenna while maintaining its compact size and adequate bandwidth.

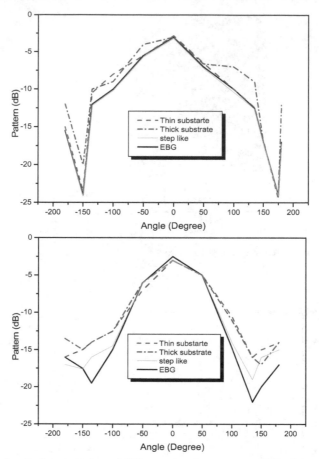

Fig. 33. Simulated radiation patterns of different patch antennas: (a) E-plane and (b) H-plane pattern.

10. Antenna design for ultra-wideand wireless communications applications

10.1 Enhancement of microstrip monopole antenna bandwidth by using EBG structures

The low profile, light weight, and low cost of manufacturing of monopole microstrip patch antennas have made them attractive candidates in many applications ranging from very high-data rate and short-range wireless communication systems, to modern radar systems. The limited bandwidth of the MPA, however, needs to be further improved to facilitate applications in UWB systems. Spiral arms shaped metallo EBG was used to increase the bandwidth of planar monopole antennas. This approach resulted in size reduction and cceptable performance from 1 to 35GHz. In this section, we present a new approach based on using a variety of shapes and sizes of embedded EBG structures. The designs of the EBG structures as well as the improvement steps in the antenna performance are described in the following sections [89].

i.Antenna Geometry and Design A low profile microstrip monopole patch antenna is proposed as shown in figure 34. Rectangular and circular geometries were simulated to optimize the performance.

Starting with a rectangular patch antenna on a rectangular ground plane and ended up with an umbrella semi-circular shape, for both the radiator and the ground plane. A semi-circular ground plan with radius R_g =15mm was used as shown in figure 34. As a last step, circular shapes for both the radiator and the ground plane with radii of R_r =12mm for the antenna and R_g=15mm for the ground plane are used as shown in figure 34. The microstrip feed line length is L_f =16mm, and width W_f =1.9mm. The antenna is printed on FR4 substrate with ε_r=4.7 and thickness 3.2mm ($0.034\lambda_o$). After optimizing the basic antenna design with circular radiator and ground plane, various designs of EBG structures are embedded to improve the bandwidth, enhance gain, and reduce the antenna size beyond that was previously reported. The proposed metallic patch antenna layout was first surrounded by an EBG lattice with circular holes with dimension computed as described in [51], hence is given by $d = \lambda_g$ where **d** is the diameter of drilled hole and λ_g is the guided wavelength. Then embedded circular and square patches in a periodical electromagnetic band-gap structure were used as shown in figure 34, and the side view is shown in figure 34. The periodic patches have radius=3mm for the circular patch and, half-square side =3mm for the square patch. The periodicity in the two cases is P=7mm, and the vias radius equals 0.25mm. These dimensions were used based on optimized simulation results. The optimization process includes embedded circular and square patches dimensions as well as antenna parameters such as gain, bandwidth, and radiation efficiency. The fabricated antenna is shown in figure 34.

ii. Simulation and Experimental Results

The simulation results are shown in figure 35 (a). Then partial rectangular ground plane is used with dimensions 30x15 mm^2. Results for this case are also shown in figure 35 (a). The partial rectangular ground plane is then converted to half circular plane to increase the bandwidth by creating adjacent staggered resonance modes. Both the radiator and the ground plane shapes are then converted to half circular shapes with radii of 12mm and 15mm, respectively. The final design for the basic antenna shows ultra wideband characteristics shown in figure 35 (b) but with bandwidth discontinuity from 7 to 10GHz and 12.5 to 17.5GHz. To remove these discontinuities, EBG structures are embedded. The effect of EBG on the antenna performance is examined first by drilling cylindrical holes each with a radius equals to 3mm and periodicity P=7mm. Then circular and square EBG patches are embedded in the substrate. Simulation results to illustrate the performance of each of the EBG structures (circular or square) as well as the case of the drilled holes are shown in figure 37. These results were obtained by calculating the transmission coefficient of a 50 Ω microstrip transmission line placed on the top of a substrate with either square or circular EBG structure or circular holes. From figure 37, it is seen that each of these embedded structures has different effect on the bandwidth of the microstrip line. The pass-band and stop-band for the embedded square are larger and deeper than that of circular embedded EBG. Embedded circular patch of radius 3mm and square patch of side length d=6mm EBG structures with periodicity P=7mm are combined with the half circular-monopole antenna as illustrated in figure 45(a) and 45(b), respectively. The comparison between simulation and measurement of reflection coefficient S_{11} of both embedded EBG are shown in figures 37 and 38, respectively. From these figures it is noted that not only improvement in the bandwidth was achieved but also the antenna size was reduced to about 34% from that of the basic half circular monopole antenna without EBG. This gives more than 60% size reduction higher than that published in [92] by 10%. The final antenna with square embedded patches has average antenna gain of about 6.5dBi. The gain value changes throughout the operating band as shown in figure 38. At lower frequencies, the gain is about 3dBi and started to increase with frequency until it reaches its maximum value of 8.5dBi at 12.5GHz. After this, it starts to decrease again with frequency. However, the antenna gain with EBG is larger than without EBG throughout the entire frequency band by an average value of about 2dBi. The antenna radiation efficiency has the highest value from 1 to 7.5GHz and then it stays between 60-70% up to 35GHz. The final antenna design has an average efficiency of 73.5% as compared to only 56.5% for the basic antenna without embedding EBG structures.

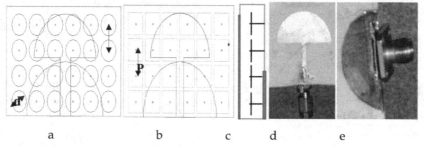

a   b   c   d   e

Fig. 34. The proposed antenna with embedded EBG (a) Circular EBG, (b) Square EBG (c) side view (d) fabricated radiator monopole antenna, and (e) ground plane on back of the substrate.

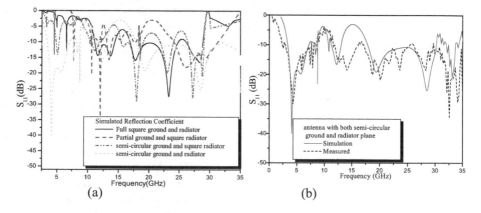

(a)   (b)

Fig. 35. (a) Simulated S_{11} vs. frequency for various design parameters and (b) comparison between simulation and measurement for the proposed.

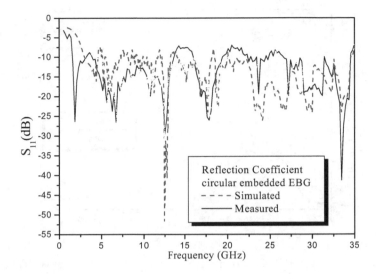

Fig. 36. Comparison between measured and simulated reflection coefficient for umbrella monopole antenna with embedded circular EBG patches.

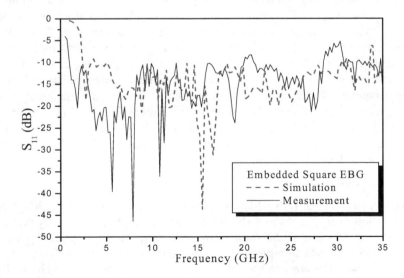

Fig. 37. Comparison between measured and simulated reflection coefficients of umbrella monopole antenna with embedded square EBG patches.

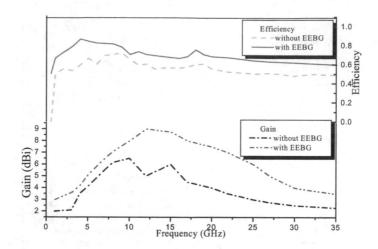

Fig. 38. Antenna gain and radiation efficiency vs. frequency for antennas with and without square EEBG. EBG structure

10.2 Ultra-wide bandwidth umbrella shaped MMPA using Spiral Artificial Magnetic Conductor (SAMC)

i.   Proposed Antenna Geometry and Design

The proposed antenna has semi-circular shape. The thickness of the substrate used is approximately $0.034\lambda_{3.3GHz}$ = (3.2 mm). The semi-circular patch of radius R_r mm, is placed on one side of dielectric substrate with relative permittivity 4.7 and tan δ =0.02. The dimension for the substrate is $L_s \times W_s$ mm². The antenna is fed by 50Ω microstrip feed-line of width 'W_f'. The semi-circular radiator is placed 'L_f' distance from one edge of the substrate. The design of the proposed antenna started from conventional shape of printed rectangular microstrip monopole. To improve the bandwidth, the radiator was modified to be semi-circular patch with radius 12mm. The antenna size is thus reduced by 22% from the original size. The dimension of FR4 substrate is 40x40 mm² but in this case discontinuities in bandwidth were found between frequency band from 1GHz to 5GHz and from 6GHz to 10GHz as shown in following sections.

ii.   Spiral AMC Design

In this section, two typical printed spiral geometries are investigated as shown in figure 39. The operation principle of the AMC surface can be simply explained by an equivalent LC circuit theory. To increase the value of the inductance, a single spiral is placed on top of the grounded substrate to replace the conventional ground plane. The parameters of the substrate remain the same as the reference conventional monopole. The width of the spiral is 1mm=$0.011\lambda_{3.3GHz}$ with gap=1mm. Two shapes of spiral AMC are used in the proposed design, one arm spiral as in figure 40(a) and four arms spiral as in figure 40(b). The two spiral shapes are applied on the antenna ground plane as shown in figure 40 (a) and 40(b) to improve the performance such as decreasing the antenna size, reducing the bandwidth discontinuities and increasing the antenna gain. After achieving the best possible results from this approach, holes were drilled to further improve the bandwidth and enhance antenna gain as shown in figure 40(c). The design of the drilled holes EBG structure is straight forward, hence, $d=\lambda_g$ where d is the diameter of drilled holes and λ_g is the guided wavelength. $d=1$mm and a periodicity $=3$mm were chosen and the fabricated antenna is shown in figure 41(a) and 41(b). The diameter of the drilled holes d is chosen to be sufficiently applied at the desired discontinuities frequencies from 6 to 10GHz. This makes the presence of the holes too small to affect the dielectric constant of the substrate.

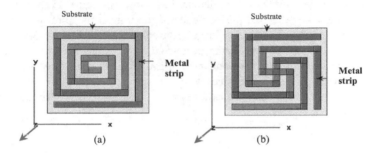

Fig. 39. The spiral shape as AMC; (a) one arm, (b) four arms.

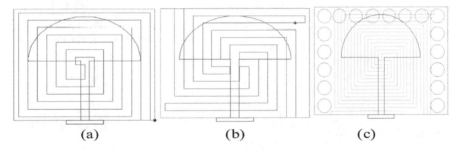

(a)   (b)   (c)

Fig. 40. Umbrella shape with spiral ground plane (a) one arm (b) four arms and (c) four arms with surrounded holes.

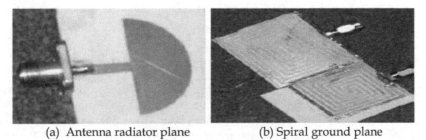

(a) Antenna radiator plane   (b) Spiral ground plane

Fig. 41. Fabricated umbrella shape antenna with spiral ground plane.

iii.   Simulation and Measurement results

In order to provide design criteria for the proposed antenna, the effects of each developing geometrical shape are studied. We started with rectangular plates for both the radiator and the ground plane. The radiator dimensions are (24x12) mm^2 and the ground plane has the dimensions of (40x40) mm^2. The simulation results are shown in figure 42. Second attempt involved converting the rectangular radiator to a semi-circular plate with radius =12mm as shown in figure 42 (same steps as in the previous sections). The electromagnetic AMC band-gap structure was used to enhance the bandwidth, increase the antenna gain and reduce the antenna size. Both the transmission line approach and reflection phase methods were used to study and redesign the performance of spiral shapes. First, the reflection coefficient and transmission coefficients S_{11} and S_{21} were calculated as shown in figure 43 for the two types of spiral. Figure 43 indicates that the performance of the four arms spiral ground plane is better than one arm spiral. Moreover, a large cross polarization was observed in many frequency bands in the one arm spiral one performance. This cross polarization resulted from the asymmetric geometry of the one arm spiral. Second, the reflection angle was calculated as shown in figure 44 for two types of spiral. Figure 44 indicates that the performance of spiral with four arms ground plane gives larger bandwidth and near from zero reflection phase.

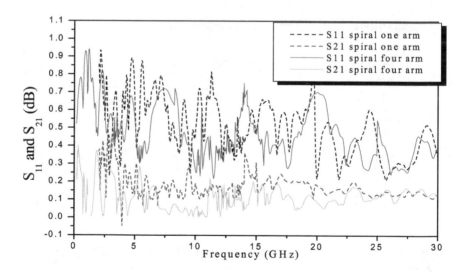

Fig. 42. Simulated results of S_{11} and S_{21} of spiral ground plane with one and four arms.

Then each of the two types of spirals was integrated as an antenna ground plane. The comparison between simulated and measured reflection coefficient for the antenna integrated with SAMC ground plane with one arm is shown in figure 45. A large frequency discontinuity was also observed in this case in the frequency band from 1 GHz up to 40 GHz. Although, the results as shown in figure 46 seem to meet the desired requirement in terms of surface compactness, the resulting high cross polarization makes it unacceptable in many applications. Due to the significant cross polarization, the single arm spiral geometry was not a good candidate for applications that require low cross polarization. Therefore, an alternative design that consists of four arms spiral is used [91].

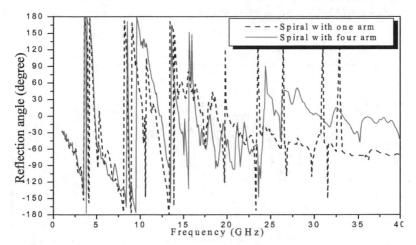

Fig. 43. The S_{11} phase of spiral AMC with one and four arms.

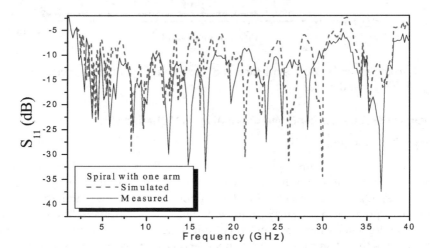

Fig. 44. Comparison between simulated and measured S_{11} of proposed antenna with one arm spiral ground plane.

The four arms spiral is then applied on an antenna ground plane. Four spiral branches, each with a 1mm =$0.011\lambda_{3.3GHz}$ width, split from the center and rotate in clockwise direction. The antenna reflection coefficient is shown in figure 46. Simulation results show that there is small discontinuity in operated antenna bandwidth between frequencies from to 0.1GHz to

1GHz, from 5.5GHz to 6.5 GHz, from 7GHz to 7.5GHz, 13.5GHz to14.5GHz In this case, the resonant frequency decreases as the number of spiral arms increases. The first resonant frequency is 49.45% lower than the reference geometry. This significant reduction in size for a single element leads to an attractive design feature for many wireless communication applications. Finally, both electromagnetic band-gap (EBG) drilled holes and artificial magnetic conductor (AMC) were merged to optimize the antenna performance as is shown in figure 42(c).

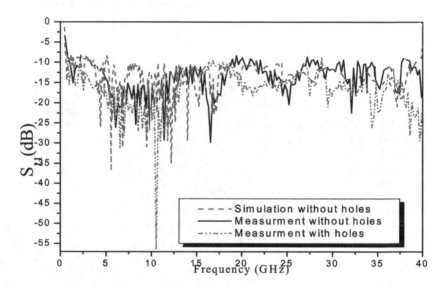

Fig. 45. The comparison between simulated and measured S_{11} of the proposed antenna with four arms spiral ground plane.

10.3 Design MMPA by using printed unequal arms V-shaped slot

i.   Design Considerations

The printed monopole antennas give very large impedance bandwidth with reasonably good radiation pattern in azimuth plane, which can be explained in two ways. To estimate the lower band-edge frequency of printed monopole antennas, the standard formulation given for monopole antenna can be used with suitable modification. The equation was worked out for the planar monopole antennas. If h is the height of substrate in cm, which is taken the same as that of an equivalent monopole and L_L in cm is the highest effective length of the V- unequal arms monopole antenna and L_f is the length of 50Ω feed line in cm, then the lower band-edge frequency is given as shown in figure 47:

$$f_L = c / \lambda = \frac{7.2}{(h + L_L + L_f)\sqrt{\varepsilon_r}} \text{GHz} \tag{4}$$

Where all dimensions h, L_L and L_f in cm.

ii. Antenna Design Methodology and Geometry

A V-shaped slot monopole with two unequal arms is proposed to achieve compact size and ultra wideband design. For the lower band (0.75GHz), two different-length arms of a V-shaped patch were used to excite two closely staggered resonant modes [84-86] as shown in figure 47. The two patches (triangular shaped patch and the electromagnetically coupled V shape are excited in the TM_{01} mode. The design achieves ultra bandwidth in lower band with antenna thickness less than $0.01\lambda_0$. Multi frequency operation is achieved by etching V slot with different spacing between feeding plate and triangular patch. It is placed on one side of dielectric substrate RT/D 6010 with relative permittivity 10.2 and tan δ =0.002. The dimension for the substrate is $L_s \times W_s$ =50x50mm². The lowest frequency of operation is f_L=0.75GHz. Monopole antenna passes by many steps to reach the final novel proposed shapes. The design of the proposed antenna started from conventional shape of microstrip monopole antenna with square substrate equal to 50 mm using rectangular ground plane and triangular plate for antenna radiator. The dimensions are optimized by choosing a 50 Ω feeding of W_f =2.5mm and L_f=16.5mm. The main objective of optimization is producing broadband antenna. The second step to improve the antenna bandwidth is modifying the ground plane to partial rectangular ground plane, the ground plane dimensions are 17.5x50 mm². Third step is using V-shaped slot with unequal arms for radiator plate as shown in figure 47. Fourth step is modifying the ground plane and optimization is used to enhance the antenna bandwidth and reduce the electrical antenna size. The triangular radiator patch with base and height W_p=17mm and L_p=15mm with arms width W_L=12mm, W_s=4mm and arms length L_L=36mm, L_s=26mm and air gap separation between triangular patch and V shaped arms g_L=2mm, g_s=3.5mm with feeding length 17.7mm. The ground plane with L_t=12mm and radius 14.4mm.

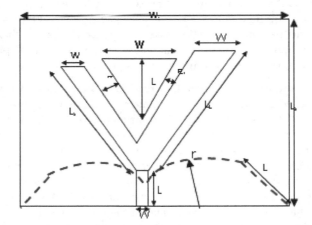

Fig. 46. The proposed antenna geometry.

$$2a = \lambda_{guide} \tag{5}$$

$$a/p = 0.8 \sim 0.9 \tag{6}$$

where a, P and λ_{guide} are side length, periodicity, and guided wavelength, respectively.

Finally, the proposed antenna is printed on metamaterial surface, first shape is using artificial magnetic conductor (AMC). As embedded spiral AMC with four arms is used to improve the impedance matching and reduce the antenna size as shown in figure 48(a) with arm width g = air gap = 2mm at height h=1.25mm. Secondly, using embedded uni-planar square EBG without via and with square side length 5mm for further improvement in antenna gain and bandwidth discontinuity as shown in figure 48(b) with side length a=5mm and periodicity P=5.5mm at height h=1.25mm. Finally, using electromagnetic band-gap structure (EBG) as embedded square EBG with square side length 4mm, periodicity 5mm and via radius =0.25mm at the same previous height as shown in figure 48(c) to enhance the antenna efficiency and gain. These dimensions were selected by recomputation in a conventional way. The EBG substrate does not interfere with the near field of the antenna, and it suppresses the surface waves, which are not included in the patch antenna design [84-86]. The design of an EBG antenna has been straightforward as shown in Eqs. 5 and 6.

iii. EBG Methodologies and Geometry

The application of EBG in printed antenna design has received significant attention recently. They have mainly been used to achieve microstrip antennas on thick, high-dielectric constant substrate with optimum performance. The approach is based on using artificial substrates made of periodic metalo-dielectric resonant implants in order to have a complete forbidden band gap around the desired antenna operative frequency. Because surface waves cannot propagate along the substrate, an increased amount of radiated power couples to space waves reducing antenna losses while increasing its gain and bandwidth. In this section, a compact AMC design that minimizes the cross polarization effect of printed spiral geometry will be presented. As the number of arms or turns order increases, the equivalent inductance increases, resulting in a lower resonant frequency. Several typical printed spiral geometries are investigated in this section and their reflection phase characteristics are reported. Due to the significant cross polarization, the single and double spiral geometries are not good candidates for applications requiring low cross polarization.

A four-arm spiral is explored to eliminate the cross polarization, as shown in figure 48(a). Each arm is rotated 90°. Therefore, this symmetrical condition guarantees the same scattering response to the x- and y-polarized incident waves. As a result, no cross polarization is observed from this structure. Second, using two dimension embedded EBG without via simplifies the fabrication process and is compatible with microwave and millimeter wave circuits. The embedded EBG surface and the AMC surface each has his own advantages. Third, a mettalo 3D-EBG with via is used: Advantages of the 3D-EBG surface is obtaining: a lower frequency and a wider bandwidth. At a given frequency, its size is smaller than the AMC design. Advantages of the uni-planar surface are obtaining a lower frequency and wider bandwidth. In addition, it is less sensitive to the incident angle and polarization.

iv. Results and Discussion

The antenna performance was investigated by both simulation and measurement as shown in figure 49. In order to provide design criteria for the proposed antenna, the effects of each developing geometrical shape are analyzed.

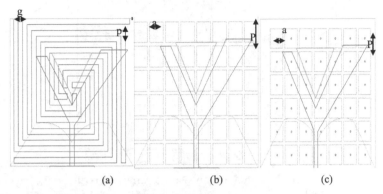

Fig. 47. The configuration of three EBG techniques.

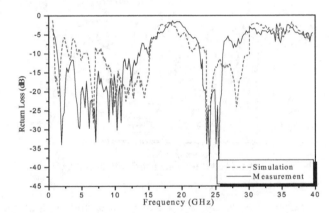

Fig. 48. Comparison between simulated and measured reflection coefficient.

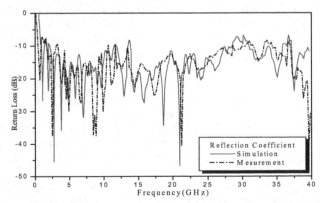

Fig. 49. Comparison between simulated and measured reflection coefficient for the proposed antenna with embedded spiral AMC

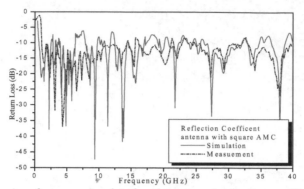

Fig. 50. Comparison between simulated and measured reflection coefficient of the proposed antenna with embedded 2D-EBG.

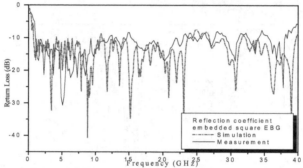

Fig. 51. Comparison between simulated and measured reflection coefficient of antenna with embedded 3D-EBG.

Metamaterials structures are used to enhance the antenna performance as bandwidth, gain and improve the reflection phase. Starting with embedded spiral AMC, the comparison between simulation and measurement of reflection coefficient is shown in figure 50, and there are bandwidth discontinuity found at frequencies from 2 to 2.5GHz, 12 to 12.5GHz, 14 to 14.5GHz and 27 to 32GHz. However, the electrical antenna size reduced to 0.6GHz with reduction 65% from original size of the proposed antenna. Secondly, embedded square uni-planar 2D-EBG were used. The comparison between simulation and measurement is shown in figure 51 and the bandwidth discontinuities occur at frequencies from 5 to 8 GHz, from 12 to 13, 17 to 18.5GHz and from 32 to 34.5GHz. Thirdly, embedded 3D-EBG is used for further improvement in antenna gain and bandwidth. The bandwidth discontinuities occur at frequencies from 11 to 12GHz and from 26 to 26.5GHz as shown in figure 52. It may be noted that not only improvement in the bandwidth is achieved but also the antenna size was reduced to about 65% from that of the proposed monopole antenna without EBG. From these figures, one can notice that there are small discrepancies between the simulated and measured results. This may be attributed due to the same reasons stated before. Antenna gain for three antenna structures are shown in figure 53. This figure shows that, the antenna with embedded 3D-EBG has the antenna gain response along the operating band with average gain of about of 13dBi followed by 2D-EBG with average gain 11dBi and spiral AMC with 9dBi. The fabricated antenna is shown in figure 54.

10.4 Co-planar boat MPA with modified ground plane by using EBG

i.   Antenna Geometry

The geometry of the proposed antenna is shown in figure 55, where an equi-lateral triangular patch with L_r=70mm is placed co-planar to a finite ground plane that has a trapezoidal shape with size of W_{g1} =30mm, W_{g2}=63mm, hieght L_g=21mm. The dielectric substrate used is FR4 with dielectric constant ε_r=4.7 and dimension 100x100mm² with thickness h = 3.2mm. The patch is proximity fed by a 50Ω microstrip line at the fundamental frequency 3.3GHz with line length and width of L_f = 63mm and W_f = 5.5 mm, respectively. The top and side views of the proposed antenna are shown in figure 55. To obtain a good impedance matching, the end of the feed line has to extend beyond the centre of the patch. Initially, different dimensions of the MPA were used in order to minimize the size of the patch antenna with maximizing the bandwidth at the same time. The antenna geometry of the whole structure should be optimized (the ground plane dimension, separation between the patch and the ground and feed line position) to obtain the best possible impedance bandwidth [87].

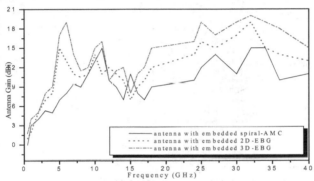

Fig. 52. Comparison of gain for the studied EBG structures.

Second part of this section, is using 2D-EBG etched in the feeding line to improve impedance matching with head square dimension a= 2mm, slot length L_d=1.25mm with width g=0.5mm and periodicity P=4mm. An etched 2D-EBG in 50Ω feed line disturbs the shield current distribution in the feed line. This disturbance can change the characteristics of the transmission line since it increases the effective capacitance and inductance of a transmission line, respectively.

Finally, using the four arms spiral AMC to reduce the antenna size by adding inductance component. A larger equivalent inductance may be realized with a larger number of spiral turns. However, as revealed in this chapter, if the unit geometry is not symmetric with respect to the polarizations of the incident waves, the AMC surface generates a high level of cross polarization. Thus, the behavior of reflection phase may not be applicable in the designated frequency band of operation. Using four arms spiral shape, and each arm rotates 90° can exactly recover itself. Therefore, this symmetrical condition guarantees the same scattering response to the x- and y-polarized incident waves. As a result, no cross polarization was observed from this structure. Therefore, with this design the compactness of geometry is achieved without generating the cross polarization level. The dimension of the spiral arm width W_d is equal to separation between arms =5mm ($0.05\lambda_g$) and the largest spiral length L_s= 80mm as shown in figure 55(b).

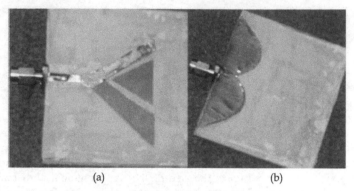

Fig. 53. The fabricated antenna (a) radiator and (b) ground plane

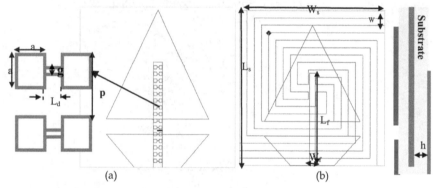

Fig. 54. Top and side view of the proposed antenna., (a) the proposed antenna with EBG elevation and (bc) side view.

ii.   Simulated and Measured Results

The antenna performance was investigated by using simulations and verified experimentally by fabricating the antenna using photolithographic techniques. In order to provide design criteria for the proposed antenna, the effects of each developing geometrical dimensions were analyzed. The effect of feeding line length was studied. There is an optimum value of the feed length which is 63mm. The simulated return loss with various ground plane width, W_{gl} studied and it is found that larger width gives a broader bandwidth as well as a lower return loss magnitude. The proposed antenna is sensitive to L_g and in fact broadband performance is obtained for L_g= 21 mm. It is known that in proximity fed patch antennas the position of the feed line under patch is important. The simulated return loss with various separations between the triangular radiator and the trapezoidal ground W_{gap} plane is also studied as shown in figure 56 where it can be seen that larger gap width gives a lower bandwidth as well as a lower return loss magnitude. However, there is an optimum separation W_{gap} which gives good reflection coefficient and bandwidth as shown in figure 57. Figure 58 shows a comparison between simulated and measured reflection coefficient of the proposed antenna with optimum dimensions. Figure 58 shows comparison between simulated and measured reflection coefficient. There is an improvement in bandwidth from 1GHz to about 40GHz with discontinuity band in the operating region. An embedded four arms spiral AMC was added at height h= 1.6mm from the ground plane to decrease the bandwidth discontinuity and cross polarization as well as increasing the antenna gain.

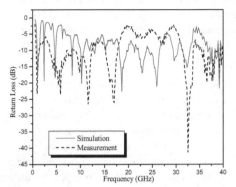

Fig. 55. Comparison between simulated and measured reflection coefficient of the antenna without EBG

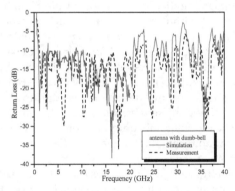

Fig. 56. Comparison between simulated and measured return loss for the antenna with dumb-bell

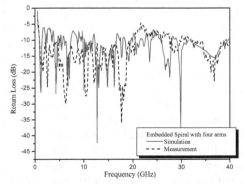

Fig. 57. Comparison between simulated and measured return loss for the proposed antenna with embedded SAMC under the antenna radiator.

11. Design of microstrip patch antenna arrays with EBG

After discussing EBG applications in single microstrip patch antennas, we now present how EBG can help to improve the performance of MPAA. The mutual coupling between array elements as an important parameter in array design. Strong mutual coupling could reduce the array efficiency and cause the scan blindness in phased array systems. [93] Therefore, the electromagnetic band gap (EBG) structures are used to reduce the coupling between array elements.

There are diverse forms of EBG structures [97], and novel designs such as EBG structures integrated with active device and multilayer EBG structures have been proposed recently. This section focuses on different types of EBG structures. Its band-gap features are revealed for two reasons, suppression of surface-wave propagation, and in-phase reflection coefficient. The feature of surface-wave suppression helps to improve antenna's performance such as increasing the antenna gain and reducing back radiation [92]. Meanwhile, the in-phase reflection feature leads to low profile antenna designs [94]. This section concentrates on the surface-wave suppression effect of the EBG structure and its application to reduce the mutual coupling of MPAA. To explore the surface-wave suppression effect, the propagating fields of an infinitesimal dipole source with and without the EBG structure are simulated and a frequency stop-band for the field propagation is identified. Furthermore, the propagating near fields at frequencies inside and outside the band gap is graphically presented for a clear understanding of the physics of the EBG structure. It is worthwhile to point out that this band-gap study is closely associated with specific antenna applications such as MPAA. Applications of MPAA on high dielectric constant substrates are of special interest due to their compact size and conformability with the monolithic microwave integrated circuit (MMIC). However, the utilization of a high dielectric constant substrate has some drawbacks. Among these are narrower bandwidths and pronounced surface waves. The bandwidth can be recovered using a thick substrate, yet this excites severe surface waves. The generation of surface waves decreases the antenna efficiency and degrades the antenna pattern. Furthermore, it increases the mutual coupling of the antenna array which causes the blind angle of a scanning array. Several methods have been proposed to reduce the effects of surface waves. One suggested approach is the synthesized substrate that lowers the effective dielectric constant of the substrate either under or around the patch. Another approach is to use a reduced surface wave patch antenna. The EBG structures are also used to improve the antenna performance. However, most researchers only study the EBG effects on a MPA element, and to the best of our knowledge there are no comprehensive results reported for antenna arrays. The mutual coupling of MPAA is parametrically investigated, including the E- and H-coupling directions, different substrate thickness, and various dielectric constants. In both coupling directions, increasing the substrate thickness will increase the mutual coupling. However, the effect of the dielectric constant on mutual coupling is different at various coupling directions. It is found that for the E-plane coupling is stronger on a high permittivity substrate than that on a low permittivity substrate. In contrast, for the H-plane coupled cases the mutual coupling is weaker on a high permittivity substrate than that on a low permittivity substrate. This difference is due to surface waves propagating along the E-plane direction. To reduce the strong mutual coupling of the E-plane coupled MPAA on a thick and high permittivity substrate, the EBG structure is inserted between antenna elements. When the EBG parameters are properly designed, the pronounced surface waves are suppressed, resulting in a low mutual coupling. This method is compared with previous methods such as cavity backed patch antennas. The EBG structure exhibits a better capability in lowering the mutual coupling than other approaches.

11.1 Different types of EBG for mutual coupling reduction

Utilization of electromagnetic band-gap (EBG) structures is becoming attractive in the electromagnetic and antenna community. In this chapter we describe three ways to improve the performance of microstrip antenna arrays by using 3D-EBG, 2D-EBG and defected ground structure DGS. At the end of this chapter as conclusion, we make comparison between the effects of these methods on the array characteristics.

11.2 Mutual coupling reduction by using the 2D-EBG structure

Surface waves are undesired because when a patch antenna radiates, a portion of total available radiated power becomes trapped along the surface of the substrate. It can extract total available power for radiation to space wave as well as there is harmonic frequency created [96]. For arrays, surface waves have a significant impact on the mutual coupling between array elements. One solution to reduce surface waves is using electromagnetic band-gap (EBG) or photonic band-gap structure (PBG). Many shapes of EBG slot have been studied for single element microstrip antenna such as circles, dumb-bells and squares. However, not many have realized in antenna arrays. It has been demonstrated that the EBG structure will lead to a reduction in the side-lobe levels and improvements in the front to back ratio and overall antenna efficiency for the radiation pattern. However, the antenna in the above mentioned references has only one patch. The unique capability of the EBG structure to reduce the mutual coupling between elements of an antenna array was demonstrated. The side lobe of the antenna with one patch is due to surface-wave diffraction at the edges of the antenna substrate. For antenna array, the side lobe is related to the pattern of the individual antenna, location of antenna in the array and the relative amplitudes of excitation. In addition, the mutual coupling between radiators affects the current distribution on the antenna and resulted in increased side lobes.

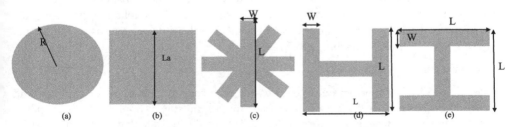

Fig. 58. The different shapes of one unit cell of 2D-EBG: (a) conventional circle, (b) conventional square, (c) star, (d) H-shape and (e) I-shape.

In this section, MPA array with three different shapes of 2D-EBG as star, H shape and I-shape slot etched on the ground plane are designed, simulated and measured. In this study, harmonic suppression and reduction of the mutual coupling effect are investigated by proposing these new shapes of 2D-EBG. The obtained results demonstrate that the 2D-EBG not only reduces the mutual coupling between the patches of antenna array, but also suppresses the second harmonic, reduces the side lobe level and gives results better than conventional 2D-EBG shapes as circle and square. It is also shown that the novel shapes of 2D-EBG on the ground plane increases the gain of the antenna array.

i.   Configurations of 2D-EBG Shapes

Three different shapes of 2D-EBG are presented, as shown in figure 59; the three shapes are compared with familiar conventional shapes as circular and square shapes 2D-EBG by using transmission line approach. The proposed EBG units are composed of several rectangular-shape slots (of length L and width W). This EBG cell can provide a cutoff frequency and attenuation pole. It is well known that an attenuation pole can be generated by a combination of the inductance and capacitance elements, which presents circuit model for the cell for all 2D-EBG structures. Here, the capacitance is provided by the transverse slot and the inductance by different shapes slots. For star shape there are four rectangular slots with $L = 5$ mm, $W = 1$ mm at angles = $0°$, $45°$, $90°$ and $135°$. The second shape, which is the H shaped slot, consists of three rectangular slots with the same previous dimensions. The third shape is the I-shaped slot, which is obtained by rotating H-shape by $90°$. The substrate with a dielectric constant of 10.2, loss tangent of 0.0019 and thickness of 2.5 mm is considered here. The microstrip feeding line on top plane has a width $W_f = 2.3$mm, corresponding to 50Ω characteristic impedance. 2D-EBG cells are etched on the ground plane with periodicity $P = 7$mm and ratio $L/p \cong 0.7$. Then the reflection and transmission coefficients (S_{11} and S_{21}) are calculated using the high frequency structure simulator (HFSS).

ii.   Antenna Array Design

Consider an ordinary antenna array with two elements, at 5.2 GHz, the dimensions of the patches are patch width $W_p = 8$mm, patch length $L_p = 7.5$mm and microstrip feed line with length L_f, $L_s =20$mm, 13mm, respectively, and the distance between the patches is $L_B = 19$ mm ($0.44\lambda_{5.2GHz}$.). It can be seen that the antenna will radiate energy at a harmonic frequency of 7.5GHz. In order to suppress such harmonics, the band-stop characteristic of the EBG structure may be used. In this section, which is simply 2D-EBG cells are etched on the ground metal sheet. The two separate array elements are also studied to measure the mutual coupling between the two patches in MIMO arrays.

iii. Results and Discussion

The response that is shown in figure 60 presents both conventional shapes as circular and square 2D-EBG as well as the three new shapes. The second step, is applying these shapes to the ground plane of two element array antenna. The results indicate that the harmonic at 7.5 GHz is indeed suppressed as well as reducing array antenna size by about 7.5% for H-shape, 8% for star-shape and for I-shape reduction in size reaches 15%. According to the characteristics of EBG, the surface wave can also be suppressed.

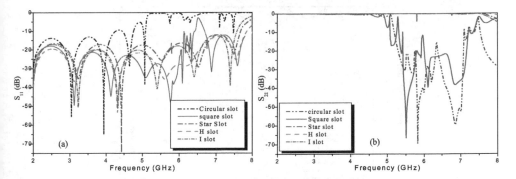

Fig. 59. (a) The reflection and (b) transmission coefficients for different 2D-EBG shapes.

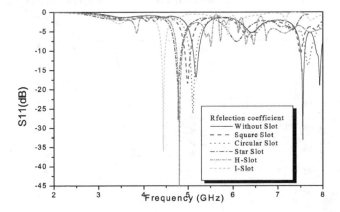

Fig. 60. The reflection coefficient for different MPAA with 2D –EBG.

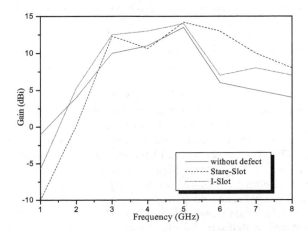

Fig. 61. The Gain response for different shapes of MPAA without and with 2D-EBG shapes.

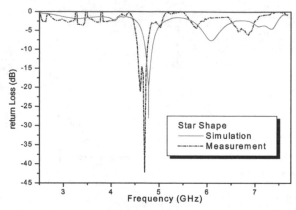

Fig. 62. Comparison between measured and simulated reflection coefficients for star shaped slot 2D-EBG.

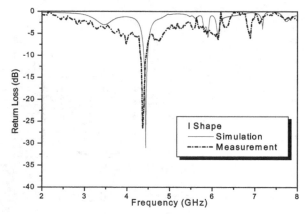

Fig. 63. Comparison between measured and simulated reflection coefficients for I shaped slot 2D-EBG.

For effective suppression of the harmonics, and for effective suppression of the surface waves, a periodic structure surrounding the patches, in addition to underneath the patches, are necessary [98]. The array performance of the conventional and the three new shapes of 2D-EBG are given in table 7. From table 7 and figure 61, one can notice that; I shape gives maximum reduction in resonant array frequency than other shapes so reduce the electrical array size, star shape gives maximum average antenna gain and minimum mutual coupling while H shape gives larger antenna bandwidth than others. By using conventional shapes as circular and square EBG, the concentration of surface current decreases but not eliminated. However, the gain of the two-element array antenna is also studied for different array antennas with and without 2D-EBG as shown in figure 62. The comparisons between measured and simulated reflection coefficients are shown in figures 63 and 64 for the star and I shape, respectively. From table 7 it notes that, at higher frequencies the antenna gain with 2D-EBG is better than that without EBG by about 9dBi maximum difference and 3dBi in average over the entire antenna band which verifies the harmonic suppression behavior.

In addition the average efficiency of the array is also studied over the operating band. The average array efficiency with conventional 2D-EBG is lower than that without by about 15% while with three new shapes of 2D-EBG is lower by about 10%.

Shape of 2D-EBG	Antenna Gain dB @5.2GHz	Mutual Coupling $|S_{21}|$	Harmonic level dB	Reflection Coefficient	F_0 GHz	BW %	Defect Geometry, Periodicity X direction /Y direction
Without	10	-16dB	-35	-17dB	5.1	3	
Square	13.5	-18.5dB	-10	-17.5dB	5	4	Side dimension 4mm, 6mm, 6mm
Circular	13	-19dB	-15	-20dB	4.8	5	Radius 2mm, 2mm, 2mm
Star	13.75	-40dB	-9	-30dB	4.75	5	Side length 1x 4mm², 6mm, 6mm
H	12.75	-20dB	-10	-40dB	5	5.1	Side length 1x5mm², 6mm, 6mm
I	11	-30dB	-7	-45dB	4	5	Side length 1x5mm², 6mm, 6mm

Table 7. The effect of different 2D-EBG shapes on the antenna performance.

11.3 Novel shapes of low mutual coupling 2X2 MPAA by using DGS

In this section, we propose new shapes of DGS structures integrated with microstrip array elements that suppress surface wave and lead to a high isolation between array elements, hence reduce the mutual coupling. Many published papers have made use of relatively complex periodic electromagnetic band-gap (EBG) structures to reduce the mutual coupling. In this section we propose improvement in E and H plane coupling by about 4dB than dumb-bell shapes presented in [95].

i.   DGS Configurations and Response

Figure 65 displays the different shapes of the DGS etched on the metallic ground plane. The different cell shapes of DGS are dumb-bell, H, E, H with inverted H and back-to-back E as shown in figure 65 (a)-(e), respectively.

The responses of these shapes are compared in figure 66 by using the conventional 50Ω microstrip transmission line approach. The transmission and reflection coefficients of the conventional dumbbell and new shapes of DGS are presented in figure 66(a) and 66(b), respectively. It is demonstrated that when an aperture etched on the ground plane, the effective inductance of the microstrip increases and the width of the connecting gap determines the shunt capacitance. The inductance and capacitance level may be independently controlled by the equivalent aperture area and gap width, respectively.

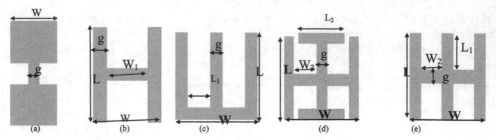

Fig. 64. Different DGS shapes (a) dumbbell, (b) H shape, (c)E, (d) H with inverted H and (e) back to back E shape.

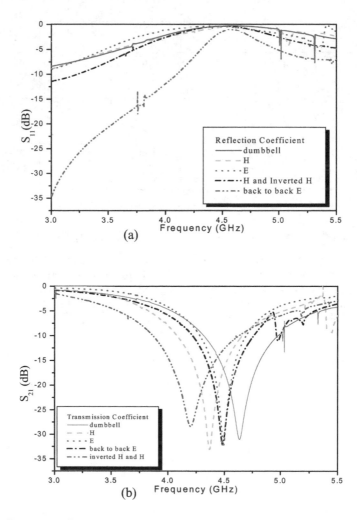

Fig. 65. (a) Reflection coefficient, and (b) Transmission coefficient for different DGS.

On the other hand, it should be noted that etching holes in the ground plane of DGS structures degrades the isolation characteristic of the ground plane thus obtaining large backward radiation pattern. However, by using these new shapes of the DGS structures, the effective apertures area are reduced thus there are improvements on the array E- and H-plane radiation patters. Figure 66 indicates that all new shapes (E, H, back-to-back E and H with inverted H) give wider stop-band filter response than conventional dumbbell shapes, hence improve the antenna characteristics.

11.3.1 Study of mutual coupling reduction

Due to high excitation of surface waves in both E *and* H-plane coupling between microstrip array patches an investigation was carried out by simulation to study the effect of the newly DGS shapes on the mutual coupling. Two substrates with thickness of 1.6 mm and 2.5mm and permittivity of 4.7 and 10.2, respectively, were used in the simulations. This difference is due to surface waves propagating along the E-plane direction, which can be easily viewed from the provided near field plots. For microstrip feed array as shown in figure 67 (a) the array element is with dimensions W_p x L_p = 16mm x 12mm and L_a=15mm. While the edge-to-edge separation is $L_B > \lambda_0/2$=30mm at a designed resonant frequency 5.25GHz and printed on FR4 substrate with dielectric constant 4.7, height 1.6mm and substrate dimensions W_s x L_s = 53mm × 53mm ($\lambda_0 \times \lambda_0$). Figure 67(b) illustrates the layout of a 4-element coaxial feed microstrip array on the defected ground plane with substrate dimensions 63mm × 63mm= ($\lambda_0 \times \lambda_0$) at a designed frequency of 4.75 GHz. The array patch is with dimensions Wp x Lp = 10.5mm x 8.5mm with edge to edge separation distance $d > \lambda_0/2$ =35mm and printed on RT/D6010 substrate with dielectric constant 10.2 and height 2.5mm. In coaxial feed, only element 2 is excited while other elements 1, 3, 4 are 50Ω terminated. The DGS dimensions for a band-gap at the resonant frequency of the antenna are optimized shapes. It consists of rectangular strips, each with dimensions L X g = 7.5mm x 1.5mm and width W=7.5mm, W=3.5mm and W=1.5mm with length L= 3mm and L= 3.5mm for E-shape, H with inverted H shape and back to back E shaped, respectively. Figure 68 shows the reflection coefficient of the microstrip line feed 2x2 MPAA at low dielectric constant FR4 substrate without and with different shapes of DGS [92]. It is observed that DGS antenna resonant frequency shifts towards lower values with respect to the conventional antenna. This small frequency shift is due to wave slowing effects of DGS. Both back-to-back E and H with inverted H give better response than other shapes. Table 8 summarizes all the MPAA characteristics without and with DGS. Figure 69 shows the comparison between simulated and measured reflection coefficient of the MPAA with back-to-back E and H with inverted H DGS. Figure 70 is a plot of all responses of the coaxial fed 2x2 MPAA at high dielectric constant RT/D6010 substrate with different shapes of DGS together with the conventional array. S_{11}, figure 70 (a), shows that DGS structure shifts down the antenna resonant frequency as compared to the conventional MPAA. The E-plane mutual coupling S_{21} are shown in figure 70 (b). The conventional antenna shows a very strong coupling of −9.63 dB due to surface waves pronounced in thick, high permittivity substrate. Since the resonant frequency 4.75GHz of the antenna falls inside the DGS band-gap, surface waves are suppressed and simulations show that mutual coupling drops to −20dB with H and inverted H-shape that is lower than the conventional by 11.28 dB. The *H*-plane mutual coupling results are shown in figure 70 (c) as S_{31}. Again, it is reduced by about 6dB compared to the conventional when using DGS shape. S_{41} which represents, orthogonal coupling, is reduced by 8dB than the conventional case as shown in figure 70 (d). Table 9, summarized all the obtained results.

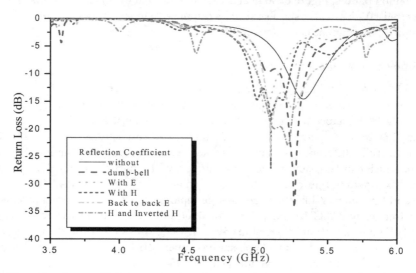

Fig. 66. The reflection coefficient comparison between antenna without and with DGS shapes for line feed array.

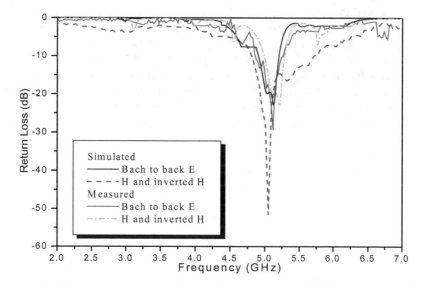

Fig. 67. Comparison between simulated and measured reflection coefficient of MPAA with defect back-to-back E and H with inverted H.

Antenna Shape	Resonant Frequency (GHz)	S_{11} (dB)	BW(MHz)	Antenna Gain (dBi)
Without Defect	5.3	-15	200	8.7
With dumb-bell	5.25	-18	250	9
With H	5.15	-17	175	9.5
With E	5.17	-30	270	9.2
Back to Back E	5.15	-28	350	10.2
H and Inverted H	5.23	-20	200	9.7

Table 8. The characteristics of the microstrip line feed array.

Antenna Shape	F_o (GHz)	Frequency response $S_{11}, S_{21}, S_{31}, S_{41}$ (dB)	BW (MHz)	Antenna Gain (dB)
Without Defect	4.75	-11, -13, -9.63, -12	150	4
With Dimple	4.7	-14, -15, -22, -25	200	4.2
With H	4.65	-15, -20, -20, -25	225	4.2
With E Shape	4.4	-14, -20, -21, -25	200	4.3
Back to Back E	4.63	-20, -20, -20, -25	250	5
H and Inverted H	4.25	-17, -20, -20, -25	200	6.5

Table 9. The characteristics of MPAA with coaxial feed

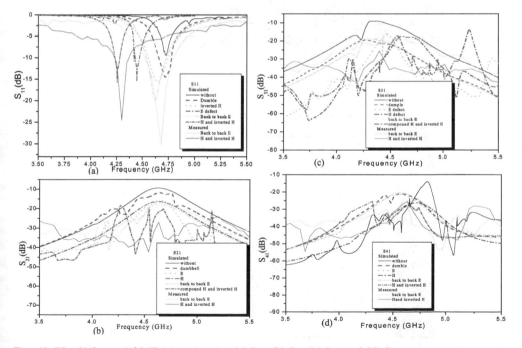

Fig. 68. The 2X2 coaxial MPAA response, (a) S_{11}, (b) S_{21}, (c) S_{31} and (d) S_{41}.

11.3.2 Influence of the different shapes of DGS on the MPAA performance

From above discussion and results in last sections, it can be concluded that:

- 1- Etching DGS on the ground plane in the middle way between the array patches improves both E-and H-plane coupling. This improvement is due to the surface waves suppression because they do not propagate along the E-Plane direction. So, the mutual coupling of MPAA is determined by both the directions of surface waves and antenna size.

- 2- The new shapes of DGS as back to back E and H with inverted H give better performance than conventional DGS shapes as dumb-bell shape.

The MPAA with back-to-back E and H with inverted H DGS shapes are fabricated using photolithographic techniques as shown in figure 71. The measured results show good agreement with simulated ones.

(a)   (b)   (c)   (d)

Fig. 69. (a) The back-to-back E, (b) H and inverted H, (c) inset line feed array and (d) coaxial feed array.

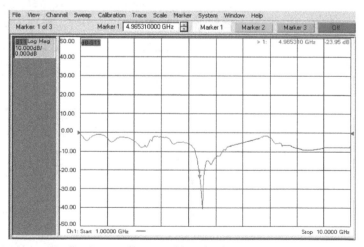

Fig. 70. The measured reflection coefficient of line feed MPAA.

11.3.3 Influence of the DGS on other antenna array characteristics

1.   Harmonic Control

Harmonic radiation is a drawback of active integrated MPAA. DGS structures are suggested to reduce the higher-order harmonics in the MPAA. The DGS antenna strongly eliminates the harmonic resonances as shown in figure 72.

2.   PIFA Array Design

For further reduction in the array size, planar inverted F antenna (PIFA) are used for array's by using patch length less than $\lambda_0/4$. The same dimensions length and width of the coaxial array patch as previous design are kept the same but shorting wall at the top end edge of each element is added. Figure 73 shows the frequency response of the 2x2 PIFA array with same excitation conditions. From the results, the newly DGS shapes improved the array by more than 5dB in S-parameters as compared with the conventional PIFA array. The reduction in array size is around 55% as compared to conventional array.

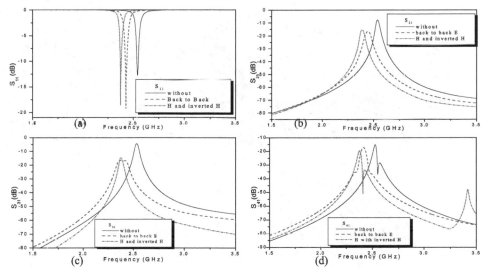

Fig. 71. The 2x2 microstrip PIFA array responses without and with back-to-back E and H with I DGS.

11.4 Ultra-wide bandwidth 2x2 MPAA by using EBG

Various types of EBG structures have been studied. In one of the first applications, a planar antenna mounted onto an EBG substrate was considered to increase the overall radiation efficiency of the device. Increasing antenna directivity was studied using an EBG structure. A compact spiral EBG structure was studied for microstrip antenna arrays. There are diverse forms of EBG structures, and novel designs such as EBG structures integrated with active device and multilayer EBG structures have been proposed recently [98].

Figure 74 shows four types of the spiral EBG structures used in this study. Comparing to other EBG structures such as dielectric rods and holes, the proposed spiral structure has a unique feature of compactness, which is important in wireless communication applications. Specifically, in this study, we investigate the use of four shapes of spiral EBG to help increase the bandwidth of a 2x2 MPAA. We choose the shape of a four-arm spiral AMC, figure 74 (b), for complete cancellation of the antenna cross polarization. This four-arm AMC is then embedded with other EBG structures to further improve the performance of the 2x2 MPAA. First it is embedded with a large four-arms spiral (LSAMC) as shown in figure 74(c), then with small spiral SSAMC patch cells with periodicity P as shown in figure 74(d), and finally with a mushroom-type EBG with spiral patches as shown in figure 74 (e). Obtained results show that the LSAMC design improves the antenna bandwidth and reduces size, while the SSAMC improves the antenna reflection phase as well as reduces the antenna array size. The embedded spiral electromagnetic band gap structure (ESEBG) in figure 74 (e), was found to improve the antenna array bandwidth and gain. Details of the specific design dimensions and the obtained simulation and experimental results are described in the following sections [98]. The concept of spiral ground plane like spiral antenna as given before in the section before,

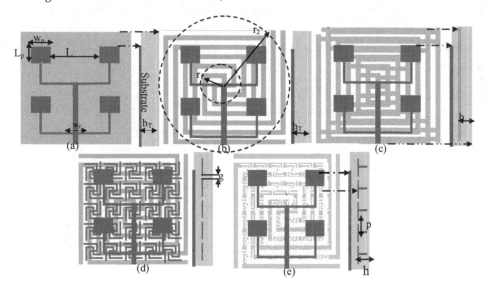

Fig. 72. Different prototype shapes of 2x2 MPAA with different EBG configurations.

11.4.1 Electromagnetic band-gap structure techniques

A typical 2x2 patch antenna array is shown in figure 74(a), where single patch length L_p =7mm, width w_p =9mm, with patch separation L_B=18mm=$0.35\lambda_{5.2GHz}$ and substrate thickness h_T = 2.5mm with conventional rectangular ground plane of dimension 50x50mm². The material of the substrate is RT/D6010 of dielectric constant ε_r =10.2. Figure 74 (b) represents the first type of EBG with four arms spiral AMC ground plane. The dimensions of the ground spiral arms, according to optimization of the transmission coefficient response of the spiral, is arm width equal to separation between arms =2mm and it rotates outward counter clockwise.

Figure 74 (c) presents prototype array antenna with second type of EBG as embedded large spiral AMC in the middle of the substrate h=1.25mm with previous ground. The dimensions of the embedded spiral is kept the same of the ground plane but less number of spiral turns and centered under the 2x2 array antenna. To add further improvement in antenna response, small cells of spiral patches with patch size dimensions 6x6 mm^2 with spiral arm width equal to gap separation=0.5mm and periodicity P=7.5mm are added at the same substrate height **h**=1.25mm with same pervious ground as shown in figure 74(d). The transmission coefficient response for this spiral structure is also shown in figure 75. From figure 75, embedded EBG gives best transmission performance followed by embedded small cells spiral with AMC then embedded large spiral AMC and the worst response is for spiral AMC in ground plane. Finally, embedded small four arms spiral patches are used as electromagnetic band-gap structure (EBG) with same pervious dimensions and with vias in the dielectric layer with radius 0.25mm.

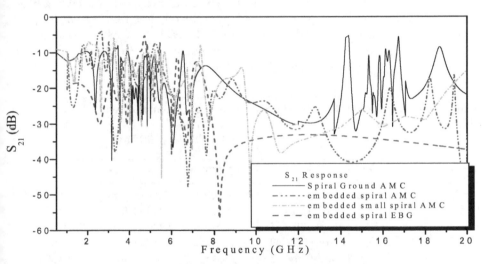

Fig. 73. The transmission response for spiral AMC ground, embedded spiral AMC, embedded small spiral AMC and embedded spiral EBG.

11.4.2 Antennas results and discussion

We started with conventional 2x2 MPAA with resonant frequency 5.2GHz. The array resonates at 5.2GHz with average gain 6.5dBi and the first harmonic appears at 7.8GHz then the conventional ground plane is reduced by four arms spiral AMC ground plane. The antenna size is reduced by 50% and the bandwidth extended from 2.5 to 19GHz with bandwidth discontinuities in the operating antenna sub-bands as shown in figure 76. Figure 77 shows the comparison between measured and simulated reflection coefficient of array antenna with spiral ground. The antenna bandwidth extended from 3GHz to 19GHz with discontinuities in bandwidth, average antenna gain is 7.8dBi. For further improvement in antenna performance as bandwidth and antenna size reduction, another embedded spiral with less number of arm turns are added at height 1.25mm from the spiral ground plane. The antenna bandwidth extended from 1.25 to 19GHz with decreased number of sub-band discontinuities as shown in figure 77 with average antenna gain 8.8dBi. Thirdly, small spiral patches are added with periodicity **P=7.5mm** to improve the bandwidth discontinuities and reflection phase especially at antenna operating frequency at 5.2GHz. To achieve optimum performance, $2a = \lambda_g$ where **a** is the side of embedded patch and λ_g is the guided wavelength. The bandwidth extended from 0.75 to 20GHz as shown in figure 77 with average antenna gain increased to 9dBi, so it is increased from original gain by about 4dBi.

Embedded electromagnetic band gap structure with four arm spiral patches are added at the same height to improve the antenna gain to 15.5dBi and good extended bandwidth without any discontinuities from 0.5 to 3.5 and from 4 to 19GHz as shown in figure 78. Figure 79 shows antenna array gain versus frequency. From this figure, all configurations give antenna gain better than the conventional array used over the entire operating band and the embedded spiral EBG gives the best performance followed by small spiral embedded AMC then the large embedded spiral AMC. To Table 10, summarizes all results of antennas array characteristics. The four different configurations were fabricated as shown in figure 80 by using the photolithographic techniques.

Antenna Ch/cs	Convention al Antenna	Antenna with SAMC Ground	Antenna with E LSAMC	Antenna with E SSAMC	Antenna ESEBG
Res. Freq.	5.25GHz	2.5GHz	2.25GHz	1.25GHz	0.75z
-10 dB BW (GHz)	Fundamenta l resonant 5.1-5.3	Extended from 1 to 3, from 3.25 to 5, from 5.2 to 9 and from 9.5 to 19GHz	Extended from 1.25 to 3, from 4 to 5, from 6 to 12 and from 12.5 to 16GHz	Extended from 1.25 to 2.5, from 2.5 to 3.5 from 4 to 5 from 5 to 7 and from10 to17 GHz	Extended from 0.5 to 3.5, and from 4 to 19.5GHz
Average Rad. Effic.	0.96	0.75	0.8	0.8	0.85
Average Antenna gain	6.5dBi	7.8dBi	8.8dBi	9dBi	10.5dBi
Fabrication	Easy	Easy	Moderate	Moderate	Hard
Reflection Phase	Bad	Medium	Good	Good	Very Good

Table 10. Characteristics of various antennas.

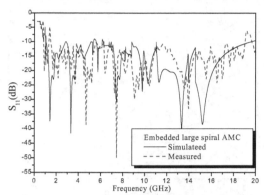

Fig. 74. The comparison between measured and simulated reflection coefficient of embedded large SAMC.

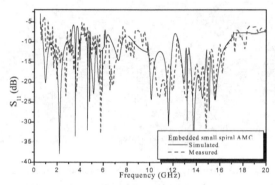

Fig. 75. Comparison between measured and simulated reflection coefficient of embedded small spiral AMC.

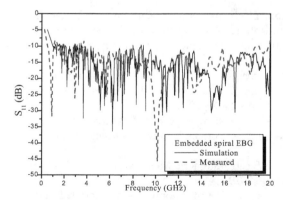

Fig. 76. Comparison between measured and simulated reflection coefficient of embedded small spiral EBG.

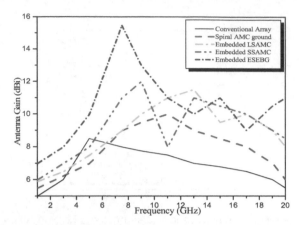

Fig. 77. Array antenna gain –vs-frequency for different configurations.

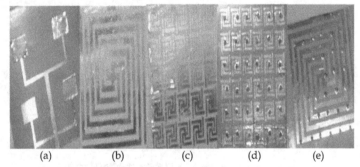

     (a)   (b)   (c)   (d)   (e)

Fig. 78. Fabricated 2x2 MPAR(a) Conventional radiator patches, (b) embedded large SAMC, (c) embedded small SAMC, (d) embedded SEBG and (e) ground plane.

12. References

[1] C. Caloz, and T. Itoh, Electromagnetic Metamaterials: Transmission Line Theory and Microwave Applications, A John Wiley & son, Inc., Canada, 2006.

[2] A. Sihvola, "Electromagnetic emergence in metamaterials. Deconstruction of terminology

[3] of complex media", *Advances in Electromagnetics of Complex Media and Metamaterials, NATO Science Series II: Mathematics, Physics and Chemistry*, Kluwer Academic Publishers, vol. 89, pp. 1 - 17, 2003

[4] A. Sihvola, "Metamaterials in electromagnetics", *Metamaterials*, vol. 1, no. 1, pp. 2 - 11.

[5] A. Sihvola, "Metamaterials: A Personal View," *Radioengineering*, vol. 18, no. 2, pp. 90 - 94, 2009.

[6] D. R. Smith, "What are Electromagnetic Metamaterials?," *Novel Electromagnetic Materials*, Retrieved Auguest, 2010.

[7] http://people.ee.duke.edu/~drsmith/about_metamaterials.html.

[8] J. G. Webster, *Metamaterial*, Wiley Encyclopedia of Electrical and Electronics Engineering, Jan. 2010.

[9] V. G. Veselago, "The Eletrodynamics of Substances with Simultaneously Negative Values of ε and μ," *Sov. Phys. Usp*, vol. 10, pp. 509 - 514, 1968.

[10] I. V. Lindell, S. A. Tretyakov, K. I. Nikoskinen, and S. Ilvonen, "BW Media–Media with Negative parameters, Capable of Supporting Backward Waves," *Microwave Opt. Tech. Lett.*, vol. 31, no.2, pp. 129-133, 2001.

[11] J. B. Pendry, A. J. Holden,W. J. Stewart, and I. Youngs, "Extremely Low Frequency Plasmons in Metallic Mesostructures," *Phys. Rev. Lett.*, vol. 76, pp. 4773 - 4776, 1996.

[12] J. B. Pendary, A. J. Holden, D. J. Robbins, and W. J. Stewart, "Magnetism from Conductors and Enhanced Nonlinear Phenomena," *IEEE Trans. Microwave Theory Tech.*, vol. 47, pp. 2075 - 2084, 1999.

[13] D. R. Smith, W. Padilla, D. C. Viers, S. C. Nemat-Nasser, and S. Schultz, "A Composite Medium with Simultaneously Negative Permeability and Permittivity,"*Phys. Rev. Lett.*, vol. 84, pp. 4184 - 4187, 2000.

[14] D. V. Smith, and J. B. Pendry, "Homogenization of Metamaterials by Field Averaging," *J. Opt. Soc, Qm.*, vol. 23, no. 3, pp. 391 - 403, 2006.

[15] J. B. Pendry, "Negative Refraction Makes a Perfect Lens," *Phys. Rev. Lett.*, vol. 85, pp. 3966 - 3969, 2000.

[16] M. C. K. Wiltshire, J. V. Hajnal, J. B. Pendry, D. J. Edwards, and C. J. Stevens, "Metamaterial Endoscope for Magnetic Field Transfer: Near Field Imaging with Magnetic Wires," *Optics Express*, vol. 11, no. 7, pp. 709 - 715, 2003.

[17] N. Katsarakis, T. Koschny, M. Kafesaki, E. N. Economou, and C. M. Soukoulis, "Electric Coupling to the Magnetic Resonance of Split Rring Resonators", *Applied Phys. lett.*, vol. 84, no. 15, pp. 2943 - 2945, 2004.

[18] D. Schurig, J. J. Mock, B. J. Justice, S. A. Cummer, J. B. Pendry, A. F. Starr, and D. R. Smith, "Metamaterial Electromagnetic Cloak at Microwave Frequencies," *Science*, vol. 314, no. 5801, pp. 977 - 980, 2006.

[19] A. Alu, and N. Engheta, "Polarizabilities and Effective Parameters for Collections of Spherical Nanoparticles Formed by Pairs of Concentric Double-Negative, Single-Negative, and/or Double-Positive Metamaterial Layers," *Journal of Applied Physics*, vol. 97, no. 9, pp. 094310 1 - 12, 2005.

[20] S. F. Mahmoud, "A New Miniaturized Annular Ring Patch Resonator Partially Loaded by A Metamaterial Ring with Negative Permeability and Permittivity," *IEEE Antennas and Wireless Propagation Letters*, vol. 3, pp. 19 - 22, 2004.

[21] A. Alu, M. G. Silveirinha, A. Salandrino, and N. Engheta, "Epsilon-Near-Zero Metamaterials and Electromagnetic Sources: Tailoring the Radiation Phase Pattern," *Physical Review B*, vol. 75, no. 15, 2007.

[22] F. Yang and Y. Rahmat-Samii, "Applications of Electromagnetic Band-Gap (EBG) Structures in Microwave Antenna Designs," Proc. of 3rd International Conference on Microwave and Millimeter Wave Technology, pp.528–31, 2002.

[23] R. Marqués and F. Medina, "An Introductory overview on right-handed metamaterials", Proc. the 27th ESA Antenna Workshop on Innovative Periodic Antennas, Spain, pp. 35-41, 2004.

[24] C. R. Simovski, P. Maagt, and I. V. Melchakova, "High impedance surfaces having resonance with respect to polarization and incident angle," *IEEE Trans. Antennas Propagat.*, Vol. 53, no. 3, pp.908–14, 2005.

[25] G. Gampala, Analysis and design of artificial magnetic conductors for X-band antenna applications, M.Sc. thesis at The University of Mississippi, 2007.

[26] M. A. Jensen, Time-Domain Finite-Difference Methods in Electromagnetics: Application to Personal Communication, Ph.D. dissertation at University of California, Los Angeles, 1994.

[27] F. Yang and Y. Rahmat-Samii, "Bent monopole antennas on EBG ground plane with reconfigurable radiation patterns," 2004 IEEE APS Int. Symp. Dig., Vol. 2, pp. 1819–1822, Monterey, CA, June 20–26, 2004.

[28] E. Yablonovitch, "Inhibited Spontaneous Emission in Solid-State Physics and Electronics," *Physical Review Letters*, vol. 58, no. 20, pp. 2059-2062, 1987.

[29] W. Barnes, T. Priest, S. Kitson, J. Sambles, "Photonic Surfaces for Surface-Plasmon Polaritons", *Phys. Rev. B*, vol. 54, pp. 6227, 1996.

[30] S. Kitson, W. Barnes, J. Sambles, "Full Photonic Band Gap for Surface Modes in the Visible", *Phys. Rev. Lett.* , vol. 77, pp. 2670, 1996.

[31] L. Brillouin, *Wave Propagation in Periodic Structures; Electric Filters and Crystal Lattices*, 2nd ed., Dover Publications, New York (1953)

[32] F. Yang, and Y. Rahmat-Samii, "Reflection Phase Characterizations of the EBG Ground Plane for Low Profile Wire Antenna Applications," *IEEE Trans. on Antennas Propag.*, vol. 51, no. 10, Oct. 2003.

[33] J. R. Sohn, H. S. Tae, J. G. Lee, and J. H. Lee, "Comparative Analysis of Four Types of High Impedance Surfaces for Low Profile Antenna Applications," *Ant. and Propagat. Society International Symposium*, vol. 1A, pp. 758 - 761, 2005.

[34] B. Wu, B. Li, T. Su, and C.-H. Liang, "Equivalent-circuit analysis and lowpass filter design of split-ring resonator DGS," Journal of Electromagnetic Waves and Applications, Vol. 20, No. 14, 1943–1953, 2006.

[35] J. Chen, Z.-B.Weng, Y.-C. Jiao, and F.-S. Zhang, "Lowpass filter design of Hilbert curve ring defected ground structure,", PIER 70, pp. 269–280, 2007.

[36] R. Sharma, T. Chakravarty, and S. Bhooshan, "Design of a Novel 3 dB Microstrip Backward Wave Coupler Using Defected Ground Structure," Progress In Electromagnetic Research, PIER 65, pp.261–273, 2006.

[37] V. Radisic, Y. Qian, R. Coccioli, and T. Itoh, "Novel 2-D photonic bandgap structure for microstrip lines," IEEE Microw. Guided Wave Lett., Vol. 8, No. 2, 69–71, Feb. 1998.

[38] F. Falcone, T. Lopetegi, and M. Sorolla, "1-D and 2-D photonic bandgap microstrip structures," Microw. Opt. Technol. Lett., Vol. 22, No. 6, 411–412, Sep. 1999.

[39] S. Zouhdi, Ari Sihvola and A. P. Vinogradov, "Applications of EBG in low profile antenna designs: what have we learned" Springer Netherlands, 2009.

[40] D. Cabric, M.S.W. Chen, D.A. Sobel, J. Yang, R.W. Brodersen, "Future wireless systems: UWB, 60GHz, and integrated Circuits Conference," pp. 793–796 cognitive radios," IEEE Proceedings of the Custom Sept. 2005.

[41] D. Elsheakh, "Electromagnetic Band-Gap (EBG) Structure for Microstrip Antenna Systems (Analysis and Design)", PhD thesis, Ain Shams University 2010.

[42] F. R. Yang, Novel Periodic Structures for Applications to Microwave Circuits, Ph.D. Dissertation, Electrical Engineering Dept., University of California, Los Angeles, 1999.

[43] M. Bradley "One-dimensional photonic band-gap structures and the analogy between optical and quantum mechanical tunneling," Eur. J. Phys, pp.108-112, 1997.

[44] N. Engheta and R. Ziolkowski, "Metamaterials: Physics and Engineering Explorations", John Wiley & Sons Inc., 2006.

[45] L. Yang, M. Fan, F. Chen, J. She and Z. Feng, "A novel compact electromagnetic-bandgap (EBG) structure and its application for microwave circuits", IEEE Trans. Microwave Theory Tech., Vol. 53, No. 1, Jan. 2005.

[46] J. Rarity, and C. Weisbuch, "In micro cavities and photonic bandgaps," Physics and Applications, Kluwer Academic Publishers, Dordrecht, 1996.

[47] Park, J. S., Kim, C. S., Kang, H. T. et al., "A novel resonant microstrip RF phase shifter using defected ground structure", 30th European Microwave Conf., pp.72-75, France, 2000.

[48] K. B. Chung, Hong, S. W. "Wavelength demultiplexers based on the superprism phenomena in photonic crystals" IEEE Trans. Antennas and Propag., vol. 81, pp. 1549-1551, Augst 2002.

[49] C. C. Chang,Y. Qian, and T. Itoh, "Analysis and applications of uniplanar compact photonic bandgap structures," PIER, vol. 41, pp. 211–235, 2003.

[50] Dahrele, S. John and Lee, "Strong Localization of Photons in Certain Disordered Dielectric Super Lattices", Physical Review Letters, vol. 58, pp. 2486 – 2489, June 1985.

[51] H. Kosaka, T. Kawashima, A. Tomita, M. Notomi, T. Tamamura, T. Sato, and S. Kawakami, "Photonic Crystals for Micro Light Wave Circuits Using Wavelength-Dependent Angular Beam Steering," Appl. Phys. Lett., vol. 74, pp. 1370–1372, Sept. 1999.

[52] D. Sievenpiper, "Chapter 11: Review of Theory, Fabrication, and Applications of High Impedance Ground Planes," in Metamaterials: Physics and Engineering Explorations, edited by N. Engheta and R. Ziolkowski, John Wiley & Sons Inc., 2006.

[53] G.-H. Li,X.-H. Jiang, and X.-M. Zhong, "A novel defected ground structure and its application to a low pass filter," Microwave and Optical Technology Letters, vol. 48, pp. 453– 456, Sep. 2006.

[54] C. S. Kim, J. S. Lim, S. Nam, K. Y. Kang, and D. Ahn, "Equivalent circuit modeling of spiral defected ground structure for microstrip line," Electron. Lett., vol. 38, pp. 1109–1120, 2002.

[55] M. Martinez-Vazquez and R. Baggen, "Characterization of printed EBG surfaces for GPS applications," IEEE Int. Workshop on Antenna Technology Small Antennas and Novel Metamaterials, pp. 5–8, March 2006.

[56] I. Garcia, "Electromagnetic band-gap (EBG) structure in antenna design for mobile communication," Project National Science Foundation, 2002-2003.

[57] Y. Ning, C. Zhining, W. Yunyi and C. M. Y. W, " A novel two-layer compact electromagnetic bandgap (EBG) structure and its applications in microwave circuits", Vol. 46 No. 4 Science in China (Series E) August 2003.

[58] X. Q. Chen, X. W. Shi, Y. C. Guo, and C. M. Xiao, " A novel dual band transmitter using microstrip defected ground structure", Progress In Electromagnetics Research, PIER 83, pp. 1–11, 2008.

[59] C. S. Kim, J. S. Park, D. Ahn, and J. B. Lim, "A novel 1-D periodic defected ground structure for planar circuits," IEEE Microwave Guided Wave Lett., vol. 10, pp. 131–133, Apr. 2000.

[60] M. Fallah-Rad and L. Shafai, "Enhanced performance of a microstrip patch antenna using high impedance EBG structure," IEEE APS Int. Symp. Dig., vol. 3, pp. 982–5, June 2003.

[61] C. C. Chiau, X. Chen, and C. G. Parini, "A microstrip patch antenna on the embedded multi-period EBG structure," Proceeding of the 6th Int. Symp. Antennas, Propagation and EM Theory, pp. 96–106, 2003.

[62] E. Yablonovitch, "Photonic band-gap structures", Journal of the Optical Society of America B, vol. 10, No. 2, pp. 283-295, Feb. 1993.

[63] H. Mosallaei and Y. Rahmat-Samii, "Periodic bandgap and effective dielectric materials in electromagnetics: characterization and applications in nanocavities and waveguides," IEEE Trans. Antennas and Propag., vol. 51, pp. 549–63, April 2003.

[64] M. Rahman and M. Stuchly, "Wide-band microstrip patch antenna with planar PBG structure," in Proc. IEEE APS Dig., vol. 2, pp.486–489, 2001,.

[65] G. Guida, A. de Lustrac, and A. Priou, "An introduction to photonic band gap (PBG) materials", PIER, vol. 41, pp.1-20, 2003.

[66] L. Qing-chun, Z. Fang-ming, and H. E Sai-ling, "A New Photonic Band-Gap Cover for a Patch Antenna with a Photonic Band-Gap Substrate", Journal of Zhejiang University Science, vol. 5, No. 3, pp. 269-273, Mar. 2004.

[67] M. A. I. El-Dahshory "Design and analysis of photonic band gap structures", Master Thesis, Cairo University, 2008.

[68] T. Sundström," Analysis of photonic crystal waveguides by the use of FDTD with regularization", Report presented to Royal Institute of Technology, 2004.

[69] A. Aminian, F. Yang, and Y. Rahmat-Samii, "Bandwidth determination for soft and hard ground planes by spectral FDTD: a unified approach in visible and surface wave regions," IEEE Trans. Antennas Propag., vol. 53, pp.18–28, January 2005.

[70] A. S. Barlevy and Y. Rahmat-Samii, "Characterization of electromagnetic band-gaps composed of multiple periodic tripods with interconnecting vias Concept, analysis, and design," IEEE Trans. Antennas Propag., vol. 49, pp. 343–353, Jun. 2001.

[71] C. C. Chiau, X. Chen, and C. G. Parini, "A multi-period EBG structure for microstrip antennas," Proceedings of 2003 ICAP, vol. 2, pp. 727–730, 2003.

[72] G. Goussetis, A. P. Feresidis, and J. C. Vardaxoglou, "FSS printed on grounded dielectric substrates resonance phenomena, AMC and EBG characteristics," IEEE APS Int. Symp. Dig., vol. 1B, pp. 644–647, July 2005.

[73] G. Goussetis, A. P. Feresidis, and J. C. Vardaxoglou, "Tailoring the AMC and EBG characteristics of periodic metallic arrays printed on grounded dielectric substrate," IEEE Trans. Antennas Propagat., vol. 54, pp. 82–9, Jun. 2006.

[74] J. R. Sohn, K. Y. Kim, and H.-S. Tae, "Comparative study on various artificial magnetic conductor for low profile", PIER 61, pp.27–37, 2006

[75] M. G. Bray and D. H. Werner, "A novel design approach for an independently tunable dual-band EBG AMC surface," IEEE APS Int. Symp. Dig., Vol. 1, pp. 289–292, June 2004.

[76] A.R. Butz "Alternative algorithm for Hilbert's space filling curve", IEEE Trans. On Computers, No. 20, pp.424-442, April 1971.

[77] P. Feresidis, A. Chauraya, G. Goussetis, J. C. Vardaxoglou and P. de Maagt, "Multiband artificial magnetic conductor surfaces", Proc. IEE Seminar on Metamaterials, for Microwave and (Sub) Millimetre Wave Applications, 24 pp. 1-4, Nov. 2003, London, UK.

[78] H. Sagan, "Space-Filling Curves," Springer-Verlag, New York, 1994.

[79] V K.J. Vinoy, K.A. Jose, V.K. Varadan, and V.V. Varadan, "Hilbert Curve Fractal Antenna: A Small Resonant Antenna for VHF/UHF Applications," Microwave & Optical Technology Letters, Vol. 29, pp. 215-219, March 2001.

[80] X. Wang Y. Hao Hall, P.S, "Dual-Band Resonances of a Patch Antenna on UC-EBG substrate," Microwave Conference Proceedings, APMC 2005.

[81] D. Nashaat, H. A. Elsadek, E. Abdallah, H. Elhenawy and M. F. Iskander "Electromagnetic Analyses and an Equivalent Circuit Model of Microstrip Patch Antenna with Rectangular Defected Ground Plane" Proceedings of IEEE international symposium on antenna and propagation AP-S, June 2009.

[82] D. Nashaat, H. A. Elsadek, E. Abdallah, H. Elhenawy and M. F. Iskander "Multiband and Miniaturized Inset Feed Microstrip Patch Antenna Using Multiple Spiral-Shaped Defect Ground Structure (DGS)" Proceedings of IEEE international symposium on antenna and propagation AP-S, June 2009.

[83] D. Nashaat, H. A. Elsadek, E. Abdallah, H. Elhenawy and M. F. Iskander "Miniaturized and Multiband Operations of Inset feed Microstrip Patch Antenna by Using Novel Shape of Defect Ground Structure (DGS) in Wireless Applications" PIERS 2009 in Moscow Progress in Electromagnetics Research Symposium, August, 2009, Moscow, RUSSIA.

[84] D. Nashaat, H. A. Elsadek, E. Abdallah, H. Elhenawy and M. F. Iskander, "Ultra-Wideband and Miniaturization of the Microstrip Monopole Patch antenna (MMPA) with Modified Ground Plane for Wireless Applications," PIERL journal, Vol. 10, pp.171-184, 2009.

[85] D. Nashaat, H. A. Elsadek, E. Abdallah, H. Elhenawy and M. F. Iskander "Ultra-Wideband Microstrip Monopole Antenna by Using Unequal Arms V- Shaped Slot printed on Metamaterial Surface" Proceedings 3rd International Congress on Advanced Electromagnetic Materials in Microwaves and Optics, London, UK, Aug., 2009.

[86] Roy, S., Foerster, J.R., Somayazulu, V.S., and D.G. Leeper. 2004. "Ultrawideband Radio Design: The Promise of High-Speed, Short-Range Wireless Connectivity." Proceedings of the IEEE. Vol. 92. pp. 295-311. Feb 2004.

[87] D. Nashaat, H. A. Elsadek, E. Abdallah, H. Elhenawy and M. F. Iskander, "Enhancement of Ultra-Wide Band Microstrip Monopole Antenna by Using Unequal Arms V-Shaped Slot Printed on Metamaterial Surface", Microwave and Optical Technology letters, Vol. 52, No. 10, pp:2203-2208, October 2010.

[88] D. Nashaat, H. A. Elsadek, E. Abdallah, H. Elhenawy and M. F. Iskander, "Ultra-Wideband Co-planar Boat Microstrip Patch with Modified Ground Plane by Using Electromagnetic Band-Gap Structure (EBG) for Wireless Communication" published in Microwave antenna Optical Technology letters, Vol 52, issue 5, pp.1159-1164, 2010.

[89] A. Yu and X. Zhang, "A low profile monopole antenna using a dumbbell EBG structure," IEEE APS Int. Symp. Dig., Vol. 2, pp. 1155–8, 20–25 June 2004.

[90] D. Nashaat, H. A. Elsadek, E.t Abdallah, H. Elhenawy and M. F. Iskander, "Enhancement of Microstrip Monopole Antenna Bandwidth by Using EBG Structures published in IEEE Antennas and Wireless Propag., Letters, vol. 8, pp.959-963, 2009.

[91] D. Nashaat, H. A. Elsadek, E. Abdallah, H. Elhenawy and M. F. Iskander, "Ultra-Wide Bandwidth Umbrella Shaped Microstrip Monopole Antenna Using Spiral Artificial Magnetic Conductor (SAMC)" IEEE Antennas and Wireless Propag., Letters, vol.8, pp.1225-1229, 2009.

[92] D. Nashaat, H. A. Elsadek, E. Abdallah, H. Elhenawy and M. F. Iskander, "Investigated New Embedded Shapes of Electromagnetic Bandgap Structures and Via Effect for Improvement Microstrip Patch Antenna Performance " PIERB journal, vol. 12, pp.90-107, 2010.

[93] D. Nashaat, H. A. Elsadek, E. Abdallah, H. Elhenawy and M. F. Iskander, "Reconfigurable Single and Multiband Inset feed Microstrip Patch Antenna for Wireless Communication Devices" PIERC journal, Vol. 12, pp.191-201, 2010.

[94] F. Yang and Y. Rahmat-Samii, Mutual coupling reduction of microstrip antennas using electromagnetic band-gap structure, Proc IEEE AP-S Dig. 2 pp. 478–481, 2001.

[95] W. Yun and Y. J Yoon, "A Wideband aperture-coupled microstrip array antenna using inverted feeding structures," IEEE Trans. Antenna and Propg., Vol. 53, No. 2, PP: 861-862, Feb. 2005.

[96] R. Gonzalo Garcia, P. de Maagt and M. Sorolla, "Enhanced patch-antenna performance by suppressing surface waves using photonic-band-gap substrate," IEEE Trans. Microwave Theory Tech., Vol. 47, No. 11, PP: 213 1-2138, November 1999.

[97] D. Nashaat, H. A. Elsadek, E. Abdallah, H. Elhenawy and M. F. Iskander, "Microstrip Array Antenna with New 2D-Electromagnetic Band Gap Structure Shapes to Reduce Harmonics and Mutual Coupling", published in PIERC journal, Vol. 20, pp.203-213, 2010.

[98] M. Salehi and A. Tavakoli, A novel low coupling microstrip antenna array design using defected ground structure, Int J Electron Commun 60, pp. 718–723, 2006.

[99] D. Nashaat, H. A. Elsadek, E. Abdallah, H. Elhenawy and M. F. Iskander,, "Low Mutual Coupling 2X2 Microstrip Patch Array Antenna by Using Novel Shapes of Defect Ground Structure (DGS)" published in Microwave antenna Optical Technology letters, Vol 52, issue 5, pp.1208-1215, 2010.

[100] D. Nashaat, H. A. Elsadek, E. Abdallah, H. Elhenawy and M. F. Iskander, "Ultra-Wide Bandwidth 2x2 Microstrip Patch Array Antenna by Using Electromagnetic Band-gap Structure (EBG)", IEEE Trans., Antenna and Propagation, vol.5, pp., 2011.

[101] R. Elliot, "On the Theory of Corrugated Plane Surfaces", IRE Trans. Ant. Prop., vol. 2, pp. 71 - 81, 1954.

[102] W. Rotman, "A Study of Single-Surface Corrugated Guides", Proc. IRE, vol. 39, pp. 952 - 959, 1951.

[103] P.-S. Kildal, "Artificially Soft and Hard Surfaces in Electromagnetics", IEEE Trans. Ant. Prop., vol. 38, pp. 1537 - 1544, 1990.

2

Investigation of Dipole Antenna Loaded with DPS and DNG Materials

Amir Jafargholi and Manouchehr Kamyab

K. N. Toosi University of Technology
Iran

1. Introduction

The increasing demands on compact multifunctional devices have necessitated the development of multi-frequency printed dipoles which can be integrated into familiar devices such as laptop computers and mobile phones. The typical difficulties encountered in designing compact antennas include narrow bandwidth, and low radiation efficiency. In order to achieve a good efficiency, considerable effort must be expended on the matching network. Other researchers have found that the bandwidth of the dipole antenna can be enhanced by loading the antenna with parallel lumped element circuits (Rogers et al., 2003). Over the last decade, increasing demands for low profile multifunctional antennas have resulted in considerable interest by the electromagnetic research community in Metamaterials (MTMs). Due to unique electromagnetic properties, MTMs have been widely considered in monopole and dipole antennas to improve their performance (Erentok et al., 2005; Erentok et al., 2008; Liu et al., 2009; Jafargholi et al., 2010). The applications of Composite Right/Left Handed (CRLH) structures to load the printed dipole have been investigated both numerically (Iizuka et al., 2006; Iizuka et al., 2007; Borja et al., 2007) and analytically (Rafaei et al., 2010). However, main drawbacks of this method are low gain and low efficiency. The use of transmission-line based MTMs to realize a tri-band monopole antenna has been recently investigated in (Zhu et al., 2010). However, the cross polarization levels of the proposed antenna in (Zhu et al., 2010) are very high. It is also known that the antenna properties can be improved by covering the metal radiating parts or filling the antenna volume. For instance, the bandwidth of the microstrip patch antenna can be significantly improved by replacing the dielectric substrate with the magneto-dielectric one (Mosallaei et al., 2007). Recently, (Erentok et al., 2008) have considered the use of Double Negative (DNG) cover to match an electrically small electric dipole antenna to free space. The effect of complex material coverings on the bandwidth of the antennas has been also investigated in (Tretyakov et al., 2004).

In this section, first, the influence of the material inclusions on the input impedance of the loaded dipoles excited by a delta function is analytically investigated. Novel and accurate analytical expressions for the input impedance of the loaded dipoles are proposed based on the mode matching technique. The boundary conditions are also enforced to obtain several simultaneous equations for the discrete modal coefficients inside the radiating region. Study of the input impedance of the whole multilayered structure is accomplished by the cascade connection of mediums as characterized by their constitutive parameters. New and accurate analytical formulas for the loaded dipole antenna are derived and successfully validated through a proper comparison with the results obtained with the commercial software CST Microwave Studio.

Moreover, a compact multiband printed dipole antenna loaded with reactive elements is proposed. The reactive loading of the dipole is inspired by the Epsilon-Negative (ENG) and DNG- MTM inclusions, which enable the loaded dipole to operate in multiband. The reactive loads are realized by two rake-shaped split ring resonators (SRRs) facing each other. Investigations reveal that the loaded dipole radiates at two or three separated bands depending on symmetrical or asymmetrical loading and load locations. The new resonance frequencies are lower than the natural resonance frequency of the conventional half wavelength dipole. In this range of frequencies, the radiation efficiency of the composite antenna is high. In order to validate the simulation results, a prototype of the proposed printed dipole is fabricated and tested. The agreement between the simulated and measured results is quite good.

2. Full-wave analysis of loaded dipole antennas using mode-matching theory

In recent years, introducing MTMs opened the way for many researcher groups to enhance the antenna performances. Due to unique electromagnetic properties, MTMs have been widely considered in monopole and dipole antennas to improve their performance (Jafargholi et al., 2010). The problem of dielectric loaded wire antenna is heretofore analyzed using numerical methods, e.g., Method of Moment (MoM) (Shams et al., 2007), Finite Difference Time Domain (FDTD) (Beggs et al., 1993), and simulations based on commercial software (Kennedy et al., 2006). However, the analytical analysis of the dielectric loaded dipole antennas has not been reported in the literature.

The novelty of this section is to introduce a mode-matching analysis of a dipole antenna loaded with material inclusions. In this section, a theoretical formulation for a multiply dielectric loaded slotted spherical antenna is proposed based on the mode-matching method, to predict the behavior of the loaded dipole. It is worth noting that the radiation pattern of a finite length small angle biconical antenna differs only slightly from the pattern of a dipole (Kraus et al., 2002). Here, since the biconical antenna can be exactly analyzed and it also reduces, in the limiting case, to a cylindrical dipole antenna (Collin et al., 1969), this structure is considered for the analytical investigations. The obtained analytical formulas confirm the general conclusions recently presented in (Shams et al., 2007; Beggs et al., 1993), regarding the effect of material inclusions on the dipole antenna performance. It is demonstrated that the inclusion influence on the input impedance of a dipole is significant only for DNG-MTM inclusions.

2.1 Field analysis

Fig. 1(a) illustrates a slotted dielectric loaded hollow conducting sphere of radius a, containing a Hertzian dipole $\bar{J} = \hat{z} J \delta(\bar{r} - \bar{r}')$, placed at the center $(r = r', \theta = 0, \phi)$, Here (r, θ, ϕ) are the spherical coordinates and δ is a delta function. The time convention of is $e^{-j\omega t}$ suppressed throughout. Due to azimuthally symmetry, the fields depend on (r, θ) and the fields are then TM waves, which can be expressed in terms of magnetic vector potentials. The total magnetic vector potential for the un-slotted sphere (First region, I) is a sum of the primary and secondary magnetic vector potentials, (Ock et al., 2009).

$$A^i(r,\theta) = \hat{z}A_z^P(r,\theta) + \hat{r}A_r^s(r,\theta) \tag{1}$$

While, the primary magnetic vector potential is a free-space Green's function as

$$A_z^P(r,\theta) = \frac{\mu_1 J}{4\pi}\frac{e^{ik_1R}}{R} \tag{2}$$

where $\hat{z}$ and $\hat{r}$ are unit vectors and $R = \sqrt{r^2 + r'^2 - 2rr'\cos\theta}$. And the secondary magnetic vector potential is

$$A_r^s(r,\theta) = \sum_{n=0}^{\infty} a_n \hat{J}_n(k_1r)P_n(\cos\theta) \tag{3}$$

where $\hat{J}_n(.)$ is the spherical Bessel function and $P_n(.)$ is the Legendre function. The coefficient is (Ock et al., 2009)

$$a_n = \frac{\mu_1 aJ}{8\pi k_1 \hat{J}'_n(k_1a)}\frac{2n+1}{n(n+1)}\int_0^\pi \Omega \frac{\partial P_n(\cos\theta)}{\partial\theta}\sin^2\theta d\theta$$
$$\Omega = \left\{(a^2 - 2r'^2 + ar'\cos\theta)(ik_1\tilde{R} - 1) + k_1^2\tilde{R}^2(a^2 - ar'\cos\theta)\right\}\frac{e^{ik_1\tilde{R}}}{\tilde{R}^5} \tag{4}$$

Now consider a slotted conducting sphere, as shown in Fig. 1(a). The total magnetic vector potential in region (I) consists of the incident A^i and scattered A_r^I potentials as

$$A_r^I(r,\theta) = \sum_{n=0}^{\infty} C_n \hat{J}_n(k_1r)P_n(\cos\theta) \tag{5}$$

Here, C_n is an unknown modal coefficient. The r-component of the magnetic vector potential in region (II, III, IV, and V) of the l-th slot is

$$A_r^\gamma(r,\theta) = \sum_{v=0}^{\infty} R_v^{l,\gamma}(\cos\theta)\left[D_v^{l,\gamma}\hat{J}_{\xi_v^{gl}}(k_\gamma r) + E_v^{l,\gamma}\hat{N}_{\xi_v^{gl}}(k_\gamma r)\right], \qquad \gamma = II, III, IV, V \tag{6}$$

Where

$$R_v^{l,\gamma}(\cos\theta) = \begin{cases} \begin{cases} Q_{\xi_v^{gl}}(\cos\theta) & v = 0 \\ Q_{\xi_v^{gl}}(\cos\alpha_2^l)P_{\xi_v^{gl}}(\cos\theta) - P_{\xi_v^{gl}}(\cos\alpha_2^l)Q_{\xi_v^{gl}}(\cos\theta) & v \geq 1 \end{cases} & II, V \\[12pt] P_{\xi_v^{gl}}(\cos\theta) + G_{\xi_v^\gamma}Q_{\xi_v^{gl}}(\cos\theta) & III, IV \end{cases} \tag{7}$$

The r-component of the magnetic vector potential in region (VI) is

$$A_r^{VI}(r,\theta) = \sum_{v=0}^{\infty} F_n \hat{H}_n^{(2)}(k_{VI}r)P_n(\cos\theta) \tag{8}$$

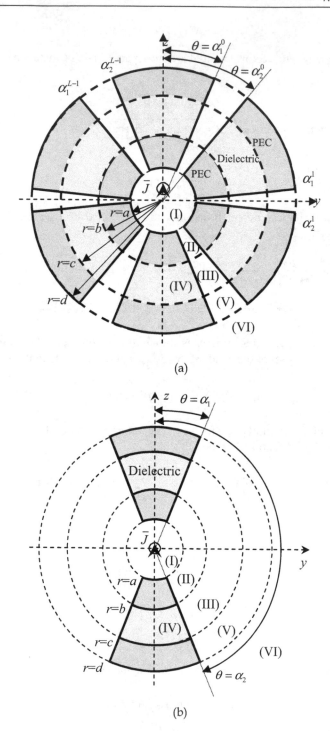

(a)

(b)

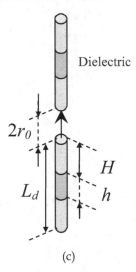

(c)

Fig. 1. (a) Multiply- (b) single slotted dielectric loaded conducting hollow sphere, and (c) dielectric loaded dipole antenna: cross-sectional view, a=0.1mm, b=2.5mm, h=$|c-b|$=0.5mm, d=5mm, r_0=0.1mm, H=2.4mm, L_d=4.9mm, and the dipole radius, r_d, is equal to 0.1mm. From (Jafargholi et. al., 2012), copyright © 2012 by the Electromagnetics, Taylor & Francis Group, LLC.

where F_n is an unknown modal coefficient and $\hat{H}_n^{(2)}(.)$ is the spherical Hankel function of the second kind. To determine the modal coefficients, we enforce the field continuities as Table 1.

Boundary	layers	Electric Fields		Magnetic Fields	Limits
1	I , II	$E_\theta^{II} = \begin{cases} E_\theta^{I} & \alpha_1^q < \theta < \alpha_2^q \\ 0 & \text{otherwise} \end{cases}$		$H_\phi^{i} + H_\phi^{I} = H_\phi^{II}$	$r=a,$ $\alpha_1^q < \theta < \alpha_2^q$
2	II , III	$E_\theta^{III} = \begin{cases} E_\theta^{II} & \alpha_1^q < \theta < \alpha_2^q \\ 0 & \text{otherwise} \end{cases}$		$H_\phi^{III} = H_\phi^{II}$	$r=b,$ $\alpha_1^q < \theta < \alpha_2^q$
3	III , V	$E_\theta^{III} = \begin{cases} E_\theta^{V} & \alpha_1^q < \theta < \alpha_2^q \\ 0 & \text{otherwise} \end{cases}$		$H_\phi^{III} = H_\phi^{V}$	$r=c,$ $\alpha_1^q < \theta < \alpha_2^q$
4	V, VI	$E_\theta^{VI} = \begin{cases} E_\theta^{V} & \alpha_1^q < \theta < \alpha_2^q \\ 0 & \text{otherwise} \end{cases}$		$H_\phi^{VI} = H_\phi^{V}$	$r=d,$ $\alpha_1^q < \theta < \alpha_2^q$
5	III , IV	$E_r^{IV} = E_r^{III}$		$H_\phi^{VI} = H_\phi^{III}$	$b<r<c,$ $\theta = \alpha_{1,2}^q$

Table 1. Boundary conditions

Applying orthogonal integrals and mathematical manipulation some can write the equations as follow.

$$C_n = -\sqrt{\frac{\mu_I \varepsilon_I}{\mu_{II} \varepsilon_{II}}} \frac{2n+1}{2n(n+1)} \frac{1}{\hat{J}'_n(k_I a)} \sum_{l=0}^{L-1}\sum_{v=0}^{\infty}\left[D_v^{l,II}\hat{J}'_{\xi v}(k_{II}a)+E_v^{l,II}\hat{N}'_{\xi v}(k_{II}a)\right]I_{vn}^{l,II} \qquad (9)$$

$$\sum_{l=0}^{L-1}\sum_{v=0}^{\infty}\left\{D_v^{l,II}\left[\hat{J}'_{\xi v}(k_{II}a)X_{uv}^{ql}-\hat{J}_{\xi v}(k_{II}a)K_v^{l,II}\delta_{ql}\delta_{uv}\right]+E_v^{l,II}\left[\hat{N}'_{\xi v}(k_{II}a)X_{uv}^{ql}-\hat{N}_{\xi v}(k_{II}a)K_v^{l,II}\delta_{ql}\delta_{uv}\right]\right\}$$
$$=-\frac{\mu_{II}Ja^2}{4\pi}L_u^q+\frac{\mu_{II}}{\mu_I}\sum_{n=0}^{\infty}a_n\hat{J}_n(k_I a)I_{un}^{q,II} \qquad (10)$$

$$\sum_{l=0}^{L-1}\sum_{v=0}^{\infty}\left[D_v^{l,\gamma}\hat{J}'_{\xi v}(k_\gamma r)+E_v^{l,\gamma}\hat{N}'_{\xi v}(k_\gamma r)\right]K_v^{l,\gamma}$$
$$=\sqrt{\frac{\mu_\gamma \varepsilon_\gamma}{\mu_{\gamma'}\varepsilon_{\gamma'}}}\sum_{l=0}^{L-1}\sum_{v'=0}^{\infty}\left[D_{v'}^{l,\gamma'}\hat{J}'_{\xi v}(k_\gamma r)+E_{v'}^{l,\gamma'}\hat{N}'_{\xi v}(k_\gamma r)\right]K_{v'}^{l,\gamma'} \qquad \begin{array}{l}\gamma=II,V\\,\gamma'=III\\r=b,c\end{array} \qquad (11)$$

$$\sum_{l=0}^{L-1}\sum_{v=0}^{\infty}\left[D_v^{l,\gamma}\hat{J}_{\xi v}(k_\gamma r)+E_v^{l,\gamma}\hat{N}_{\xi v}(k_\gamma r)\right]K_v^{l,\gamma}$$
$$=\frac{\mu_\gamma}{\mu_{\gamma'}}\sum_{l=0}^{L-1}\sum_{v'=0}^{\infty}\left[D_{v'}^{l,\gamma'}\hat{J}_{\xi v}(k_\gamma r)+E_{v'}^{l,\gamma'}\hat{N}_{\xi v}(k_\gamma r)\right]K_{v'}^{l,\gamma'} \qquad \begin{array}{l}\gamma=II,V\\,\gamma'=III\\r=b,c\end{array} \qquad (12)$$

$$\sum_{v=0}^{\infty}\xi_v^l(\xi_v^l+1)\left[D_v^{l,III}U_v+E_v^{l,III}U_{vw}\right]R_v^{l,III}(\cos\theta_0)=$$
$$\frac{\mu_{III}\varepsilon_{III}}{\mu_{IV}\varepsilon_{IV}}\sum_{v'=0}^{\infty}\xi_{v'}^l(\xi_{v'}^l+1)\left[D_{v'}^{l,IV}U_{v'v}+E_{v'}^{l,IV}U_{v'w}\right]R_{v'}^{l,IV}(\cos\theta_0), \qquad \theta_0=\alpha_1^l,\alpha_2^l \qquad (13)$$

$$\sum_{l=0}^{L-1}\sum_{v=0}^{\infty}\left\{D_v^{l,V}\left[\hat{J}_{\xi v}(k_V d)\Psi_{uv}^{ql}-\hat{J}_{\xi v}(k_V d)K_v^{l,V}\delta_{ql}\delta_{uv}\right]+E_v^{l,V}\left[\hat{N}'_{\xi v}(k_V d)\Psi_{uv}^{ql}-\hat{N}_{\xi v}(k_V d)K_v^{l,V}\delta_{ql}\delta_{uv}\right]\right\}=0 \quad (14)$$

$$F_n=-\sqrt{\frac{\mu_{VI}\varepsilon_{VI}}{\mu_V \varepsilon_V}}\frac{2n+1}{2n(n+1)}\frac{1}{\hat{H}_n''^{(1)}(k_{VI}d)}\sum_{l=0}^{L-1}\sum_{v=0}^{\infty}\left[D_v^{l,V}\hat{J}_{\xi v}(k_V d)+E_v^{l,V}\hat{N}'_{\xi v}(k_V d)\right]I_{vn}^{l,V} \qquad (15)$$

The required definitions are illustrated in the appendix. For a single slot configuration (biconical antenna loaded with a dielectric, Fig. 1(b), due to the magnetic field boundary condition between region III and IV, $R_v^{l,\gamma}(\cos\theta)$ has been simplified as

$$R_v^{1,\gamma}(\cos\theta)=\begin{cases}\begin{cases}Q_v(\cos\theta) & v=0\\Q_v(\cos\alpha_2)P_v(\cos\theta)-P_v(\cos\alpha_2)Q_v(\cos\theta) & v\geq1\end{cases} & II,V\\[2em]P_v(\cos\theta)+\left(\dfrac{\mu_{III}}{\mu_{IV}}-1\right)\left[\dfrac{P_v'(\cos\alpha_1)-P_v'(\cos\alpha_2)}{Q_v'(\cos\alpha_1)-Q_v'(\cos\alpha_2)}\right]Q_v(\cos\theta) & III\\[2em]P_v(\cos\theta) & IV\end{cases} \qquad (16)$$

Finally, the unknown coefficients are

$$C_n, \ D_v^{II}, \ E_v^{II}, \ D_v^{III}, \ E_v^{III}, \ D_v^{IV}, \ E_v^{IV}, \ D_v^{V}, \ E_v^{V}, \ F_n$$

2.2 Numerical analysis

From the formulas presented in the previous section it is straightforward to write short programs that illustrate the difference between the different types of material inclusions. To this aim, the cone angle of the biconical antenna is selected to be as small as possible, e.g., $2a_1=2.5$ degree. To clear this selection, it is should be noted that, based on (Collin et al., 1969), it is well known that the input impedance of a biconical antenna changes significantly by changing cone angle. Hence the input impedance of a biconical antenna is investigated with regards to its cone angle. The inverse radiation impedance Z_v of biconical antennas, for the small feed gap condition ($k_1a \ll 1$) is given by (Ock et al., 2009; Saoudy et al., 1990)

$$Z_v = \frac{j\eta_2 \sin\alpha_1 \ln\left(\cot\dfrac{\alpha_1}{2}\right)\displaystyle\sum_{v=0}^{\infty}\left[D_v^{II}\hat{J}_{\xi_v}\left(k_{II}b\right)+E_v^{II}\hat{N}_{\xi_v}\left(k_{II}b\right)\right]\dfrac{\partial R_{\xi_v}\left(\cos\theta\right)}{\partial\theta}\Bigg|_{\theta=\alpha_1}}{\pi\left[D_0^{II}\hat{J}_0'\left(k_{II}b\right)+E_0^{II}\hat{N}_0'\left(k_{II}b\right)\right]} \qquad (17)$$

The analytic simulations have been compared with CST simulation results of an equivalent dipole antenna (radius, r_d). The results have been presented in Fig. 2. According to these results for the antenna radius $r_d <0.01\lambda$ ($\approx$biconical antenna $2a_1 \leq 3.4$ degree, with regards to $f=25$GHz as main frequency) the loaded dipole may be considered as a limit case of a loaded biconical antenna (the approximation meet numerical simulations with good agreement). The simulation parameters have been considered as: $a=0.1$mm, $b=2.5$mm, $h=|c\text{-}b|=0.5$mm, $d=5$mm, $r_0=0.1$mm, $H=2.4$mm, $L_d=4.9$mm, and the dipole antenna filled with DPS material inclusions, ($\varepsilon_r=2.2$ and $\mu_r=1$). It is should be noted that for the radius $0.01\lambda<r_d<0.02\lambda$, the antenna input impedance has been extracted approximately; and larger values cause significant errors in impedance computations.

a.   Dielectric-Covered Biconical Antennas

To validate the proposed method, it is useful to consider a conventional covered biconical antenna (Fig. 3) as first limiting case. The input impedance of a thin biconical antenna embedded in dielectric material has been derived by (Tai et al., 1958). A slightly more general expression applicable to a biconical antenna embedded in a lossless material of arbitrary permeability and permittivity has been given by (Polk et al., 1959). Assuming $L=1$, the region III fills by PEC, and $k_1a \ll 1$; the slotted conducting sphere becomes a biconical antenna, as shown in Fig. 3. In Fig. 4, the effects of the numbers of modes in computation convergence have been depicted. It is clear that good convergence has been achieved.

In Fig. 5, the analytic results for the impedance of a biconical antenna have been compared with CST simulation results. As it is stated before, the antenna cone angle has been chosen as $2a_1=2.5$ degree. According to this figure, a good agreement has been achieved between analytic and numeric simulations.

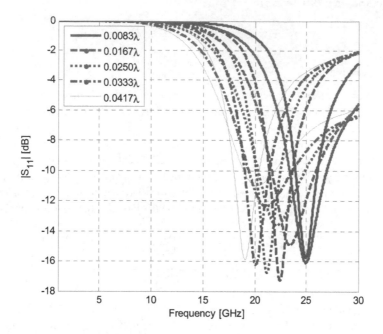

Fig. 2. The $|S_{11}|$ [dB] for a DPS-loaded dipole antenna: analytical (Blue) against numerical results (Red). Analytical results are obtained using proposed analytical expressions; while the numerical results are extracted using CST software. From (Jafargholi et. al., 2012), copyright © 2012 by the Electromagnetics, Taylor & Francis Group, LLC.

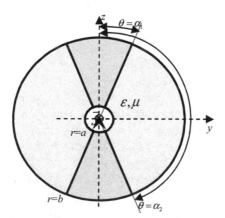

Fig. 3. Dielectric-covered biconical antenna: cross-sectional view, a=0.1mm, b=5mm, $\varepsilon_r = \mu_r = 1$. From (Jafargholi et. al., 2012), copyright © 2012 by the Electromagnetics, Taylor & Francis Group, LLC.

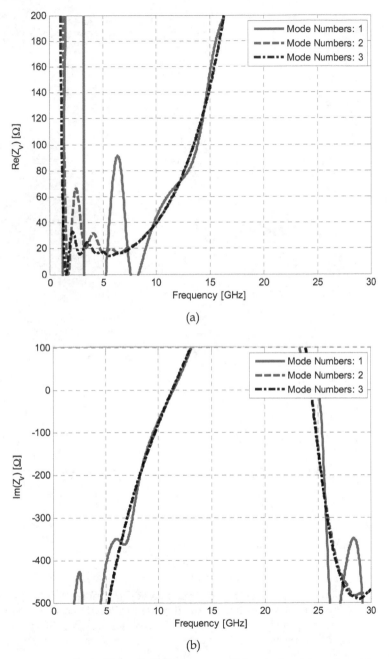

Fig. 4. Convergence analysis of the dielectric- covered biconical antenna, input impedance (a) real, and (b) imaginary parts. From (Jafargholi et. al., 2012), copyright © 2012 by the Electromagnetics, Taylor & Francis Group, LLC.

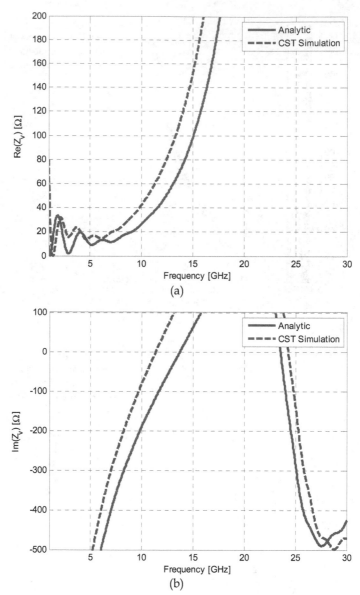

Fig. 5. Input impedance of the dielectric-covered biconical antenna: analytical against numerical results. Analytical results are obtained using proposed analytical expressions; numerical results are computed by CST software; (a) real, and (b) imaginary parts. From (Jafargholi et. al., 2012), copyright © 2012 by the Electromagnetics, Taylor & Francis Group, LLC.

In Fig. 6, the analytic result for impedance of dielectric covered biconical antenna versus dielectric material has been illustrated. The antennas parameters are b=5mm, $2a_1$=2.5 degree,

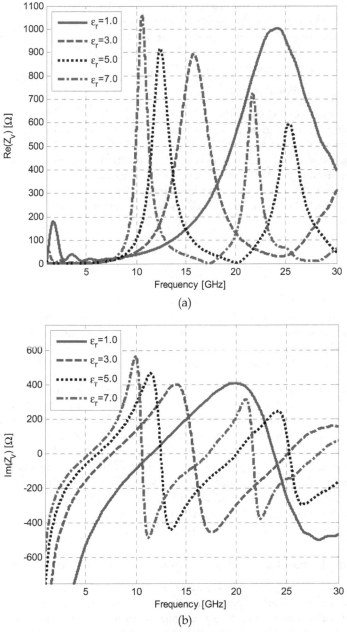

Fig. 6. Input impedance of the dielectric-loaded biconical antenna: analytical against numerical results. Analytical results are obtained using proposed analytical expressions; numerical results are computed by CST software; (a) real, and (b) imaginary parts. From (Jafargholi et. al., 2012), copyright © 2012 by the Electromagnetics, Taylor & Francis Group, LLC.

a=0.1mm. It seems clearly that the antenna input impedance affects significantly with dielectric material. The insertion of dielectric material into the biconical antenna structure causes increased frequency dependence of the antenna. The larger material permittivity, the more rapid is the variation of the input impedance. Similar results have been obtained by (Saoudy et al., 1990; Tai et al., 1958; Polk et al., 1959).

b.   *Dielectric Loaded Biconical Antenna*

In order to demonstrate the capability of the MTM loading to realize a miniaturized antenna, two examples are studied here. The first one is a dipole antenna filled with Double Positive (DPS) material inclusions, (ε_r=2.2 and μ_r=1). A DNG-loaded dipole antenna, whose parameters are labeled in Fig. 1(c), is also studied. Here, the Drude model (Jafargholi et al., 2011) is used to simulate the MTM inclusions, since it can yield a negative real part of the permittivity/permeability over a wide frequency range. For the DNG inclusions, both μ and ε obey the Drude model (with plasma frequency ω_p=15×10^{10} rad/s and collision frequency f_c =0.01GHz) as bellow

$$\xi_r(\omega) = \xi_\infty - \frac{\omega_p^2}{\omega(\omega - iv_c)} \qquad \xi \in \{\varepsilon, \mu\} \tag{18}$$

In Fig. 7, the effects of the numbers of modes in computation convergence have been presented. Again, it is clear that good convergence has been obtained. The computation time of the analytic model is about 5 minutes for all frequency points compared to several hours using CST over a frequency range of 0 to 30GHz, while this time increased in CST for higher permittivity and permeability materials. (For a 3.2GHz dual core CPU with 2GByte RAM).

The analytical results for the input impedance (both real and imaginary parts) of the DPS- and DNG-loaded dipole antennas are presented in Fig. 8. As a reference, the simulated input impedance of equivalent DPS- and DNG- loaded dipoles are also plotted in this figure. As can be seen in this figure, the analytical results for the input impedance of the loaded dipoles are in good agreement with the CST simulation results. Simulations show that for the dipole antenna loaded with DNG-inclusions, an additional resonance frequency is introduced at the frequencies lower than the antenna resonant frequency where the antenna radiates an omnidirectional radiation pattern. In contrast, for the dipoles loaded with DPS-inclusions, changing DPS locations on the antenna arms causes no resonances at frequencies lower than the main resonant frequency.

3. A compact multi-band printed dipole antenna loaded with single-cell MTM

Now, the effect of material inclusions embedded in a simple dipole antenna has been investigated. The numerical investigations result in some general conclusions regarding the effect of material inclusions on the dipole antenna performance. It is demonstrated that in contrast to the DPS and Mu-Negative (MNG) MTMs, ENG- and DNG-MTM inclusions can provide multi-band performance. To practically realize this method, a compact multiband printed dipole antenna is designed using reactive loading, which is inspired by ENG-MTM inclusions. To this aim, a novel printed MTM element is proposed and successfully tested. The proposed MTM cell shows ENG behavior at around the antenna operating frequency.

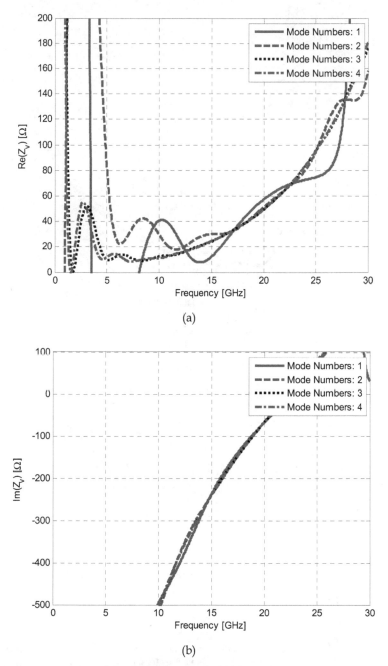

(a)

(b)

Fig. 7. Convergence analysis of the dielectric- loaded biconical antenna, input impedance (a) real, and (b) imaginary parts. From (Jafargholi et. al., 2012), copyright © 2012 by the Electromagnetics, Taylor & Francis Group, LLC.

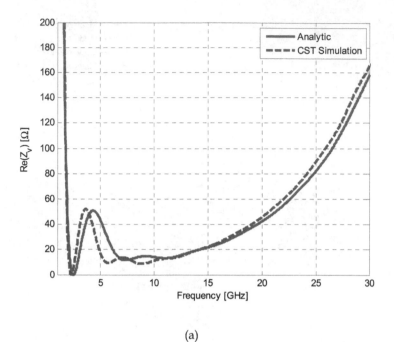

(a)

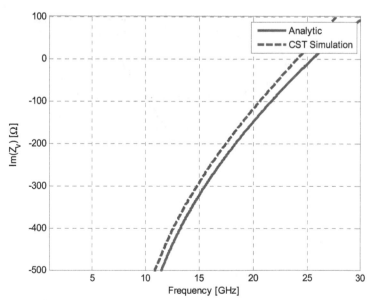

(b)

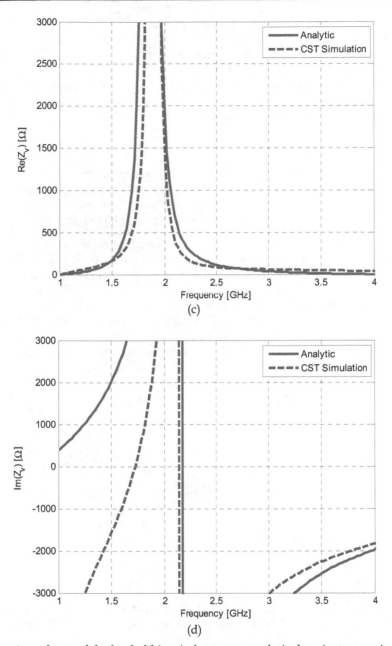

Fig. 8. Input impedance of the loaded biconical antenna: analytical against numerical results. Analytical results are obtained using proposed analytical expressions; numerical results are computed by CST software; (a) real, and (b) imaginary parts for dielectric materials and (c) real, and (d) imaginary parts for DNG metamaterials. From (Jafargholi et. al., 2012), copyright © 2012 by the Electromagnetics, Taylor & Francis Group, LLC.

The dimensions of the proposed MTM cell is optimized to meet the specifications of the mobile bands (890.2MHz–914.8MHz, and 1710MHz–1784MHz) while maintaining its compact size. The antenna radiation efficiency at the first resonance frequency is significantly higher than those reported for other miniaturized printed dipoles in the literature (Iizuka et al., 2006; Iizuka et al., 2007; Borja et al., 2007; Rafaei et al., 2010). It is worthwhile to point out here that the subject of single-cell MTM loading is not new and has been studied by other authors (Zhu et al., 2010).

3.1 A dipole antenna loaded with MTM inclusion

It is known that the resonance frequencies of an original monopole/dipole are harmonics of the main resonant frequency ω_1. However, ominidirectional radiation pattern distortion and low directivity are two major disadvantages associated with monopole/dipole antenna resonating at higher order harmonics ($\omega_m > \omega_1$) (Balanis, 1989; Jafargholi et al., 2010).

In this section, a simple and intuitive rule for determining the beneficial filling material type for dipole antennas has been introduced. A dipole antenna loaded with cylindrical dispersive MTM inclusions is shown in Fig. 9. It is assumed that the MTM inclusions are embedded in the both arms of the dipole. Here, the Drude model (Engheta et al., 2006) is used to simulate the MTM inclusions, since it can yield a negative real part of the permittivity/permeability over a wide frequency range. Depending on the MTM type either μ or ε (or both) obey the Drude model (with plasma frequency $\omega_p = 1.8 \times 10^{10}$ rad/s and collision frequency $f_c = 0.2$GHz) and are equal to one otherwise. The distance from the location of the MTM inclusions to the feed point is denoted as d_{MIF}.

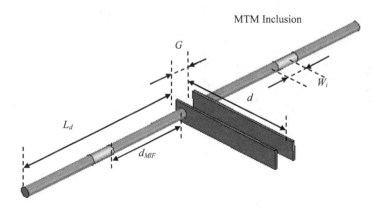

Fig. 9. An ideal model of MTM loaded dipole: L_d=120mm, W_i=2.5mm, G=5mm, d=27mm. From (Rafaei et al., 2011), copyright © 2011 by the IET Microwaves, Antennas & Propagation.

The behaviors of the loaded dipole as a function of the MTM type and the distance of the MTM inclusions from the antenna feed point, d_{MIF}, have been studied. Fig. 10 shows the antenna reflection coefficient for the dipoles loaded with DPS-, MNG-, DNG-, and ENG-inclusions, with d_{MIF} as a parameter. As the ENG- or DNG- inclusions are added, the antenna resonant behavior changes.

It can be concluded from Fig. 10 that for the dipole antenna loaded with DNG- or ENG-inclusions, an additional resonance frequency is introduced at the frequencies lower than the antenna resonant frequency where the antenna radiates an omnidirectional radiation pattern. In contrast, for the dipoles loaded with DPS- or MNG-inclusions, changing DPS/MNG locations on the antenna arms causes no resonances at frequencies lower than the main resonant frequency, as shown in Fig. 10(a,b).

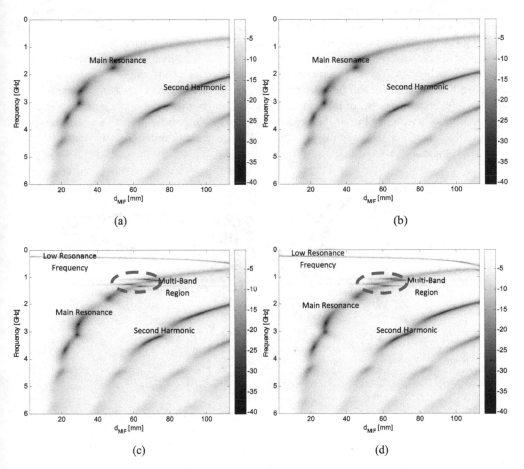

Fig. 10. CST simulation results for $|S_{11}|$ [dB], versus location of (a) DPS-, (b) MNG-, (c) DNG- and (d) ENG-inclusions. From (Rafaei et al., 2011), copyright © 2011 by the IET Microwaves, Antennas & Propagation.

As the distance between the ENG-/DNG-inclusions and the feed point is increased, the main resonant frequency decreases while the low resonant frequency is almost unchanged. This feature provides the ability to choose the second resonance frequency arbitrarily based on provision dictated by application. And thus the frequency ratio between these two frequencies can be readily controlled by adjusting the inclusion locations.

In addition, for the case of the dipoles loaded with DNG-/ENG-blocks and 50mm<d_{MIF}<75mm, more than one resonance is introduced at around the antenna main resonant frequency where the antenna radiates omnidirectional radiation patterns, as shown in Fig. 10(c,d).

To make the concept more clear, three DNG loaded dipoles are designed and simulated. The reflection coefficient results for the dipole antennas loaded with different DNG blocks and different d_{MIF} are shown in Fig. 11. For comparison purposes, the reflection coefficient of an unloaded dipole antenna is also presented in Fig. 11. As can be seen, all the antennas have multi-resonance behavior. The first frequency bands of the proposed loaded dipoles are narrow. This narrow frequency bands are the direct consequence of the resonant nature of the MTM inclusions. The gain, efficiency, and bandwidth of the three loaded dipoles are compared in Table 2. For the first design, the antenna bandwidth at first resonance is quite good but its gain is low. In contrast, for the second design, the antenna has a high gain at the first resonance frequency but at the expense of a narrower bandwidth. As a result, the type of the DNG-inclusion is a result of a trade-off between the antenna radiation efficiency (gain) and bandwidth, such as design III.

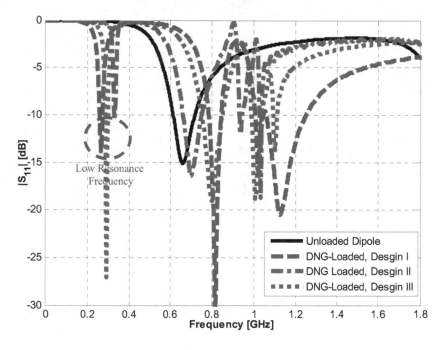

Fig. 11. Reflection coefficient results for dipole antennas loaded with different DNG blocks; Design I: d_{MIF}=72mm and Drude model with ω_p=1.8×10^{10}rad/s, and fc =0.2GHz, Design II: d_{MIF}=100mm, ω_p=1.8×10^{10}rad/s and f_c =0.01GHz, and Design III: d_{MIF}=85mm, ω_p=1.8×10^{10}rad/s and f_c =0.1GHz. As a reference, an unloaded dipole antenna is also simulated. From (Rafaei et al., 2011), copyright © 2011 by the IET Microwaves, Antennas & Propagation.

Design Type	Design I				Design II				Design III			
	f_{r1}	f_{r2}	f_{r3}	f_{r4}	f_{r1}	f_{r2}	f_{r3}	f_{r4}	f_{r1}	f_{r2}	f_{r3}	f_{r4}
Center Frequency (GHz)	0.26	0.8	1.0	1.1	0.33	0.7	0.93	1.02	0.29	0.8	1.0	1.1
Gain (dBi)	-7.2	1.25	0.0	2.1	1.3	2.1	1.9	3.4	-3.5	2.0	0.5	2.2
Efficiency (%)	12.2	78	65	91	84	99	100	100	25	91	80	95
Bandwidth (%)	5.6	7.5	5.4	21.8	1.1	13.1	2.4	1.5	4.5	12.1	5	1.8

Table 2. Gain, efficiency, and bandwidth characteristics of the dipole antenna loaded with different DNG inclusions

The behaviors of the loaded dipole as a function of the plasma frequency for d_{MIF}=72mm, have been also studied. Fig. 12 shows the antenna reflection coefficient for the dipoles loaded with ENG-, DNG-, and MNG-inclusions. It can be concluded from Fig. 12 that for the dipole antennas loaded with DNG- or ENG-inclusions, an additional resonance frequency is introduced at the frequencies lower than the antenna main resonant frequency. In addition, as can be seen from Fig. 12(a,b), in the DNG-loaded case, an additional resonance has been appeared, especially for higher values of plasma frequency, as compared to the ENG-loaded dipole. In contrast, for the dipoles loaded with MNG-inclusions, changing plasma frequency causes no resonances at frequencies lower than the main resonant frequency, as shown in Fig. 12(c).

In Fig. 13, the effect of permittivity and permeability of the DPS inclusions on the resonance frequency of the dipole antenna has been studied. As can be seen, increasing the permittivity of the loaded DPS results in a dual band operation in which the frequency separation ratio increases as the permittivity increases. However, for the dipoles loaded with magnetic inclusions, changing permeability causes no resonance frequency change.

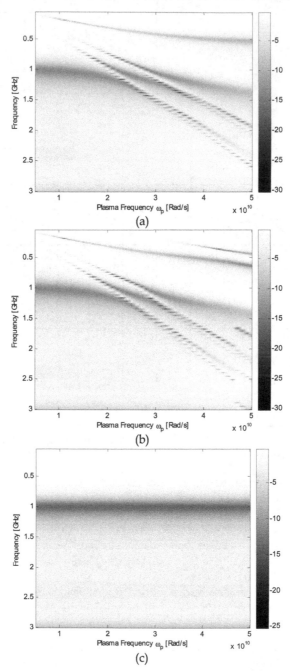

Fig. 12. CST simulation results for $|S_{11}|$ [dB] versus plasma frequency, d_{MIF}=72mm, (a) ENG-, (b) DNG-, (c) MNG-inclusions. From (Jafargholi et. al., 2012), copyright © 2012 by the ACES Journal.

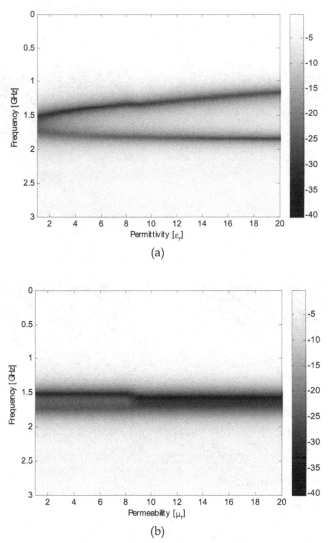

Fig. 13. CST simulation results for $|S_{11}|$ [dB], d_{MIF}=45mm, versus material (a) permittivity, μ_r=1, and (b) permeability, ε_r=1. From (Jafargholi et. al., 2012), copyright © 2012 by the ACES Journal.

The behaviors of the loaded dipole as a function of the inclusion width for d_{MIF}=45mm, have been also studied. Fig. 14 shows the antenna reflection coefficient for the dipoles loaded with DPS-, ENG-, DNG-, and MNG-inclusions. It can be concluded from Fig. 14(c,d) that, for the dipole antennas loaded with DNG-/ENG- inclusions, the additional resonance frequency is significantly affected by the width of the inclusions. In contrast, for the dipoles loaded with DPS- and MNG-inclusions, changing inclusion widths causes no resonances at frequencies lower than the main resonant frequency, as shown in Fig. 14(a,b).

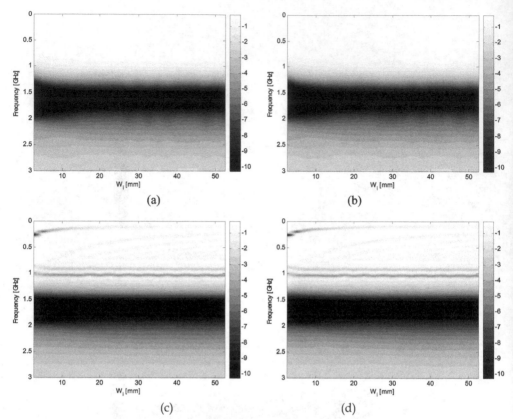

Fig. 14. CST simulation results for $|S_{11}|$ [dB], d_{MIF}=45mm, versus width of (a) DPS-,
(b) MNG-, (c) DNG- and (d) ENG-inclusions. From (Jafargholi et. al., 2012), copyright © 2012
by the ACES Journal.

4. Simulations and realization

In the previous section, it was revealed that the use of the ENG- and DNG-inclusions has led
to a multi-resonance behavior. In this section, a new printed MTM cell is introduced to
realize the ENG-inclusions. Fig. 15 shows a schematic of the proposed MTM cell along with
its design parameters. The proposed MTM cell is printed on a FR4 substrate with a thickness
of 0.8mm and a dielectric constant of 4.4. An important feature of the proposed MTM is that
it offers more degrees of freedom than conventional MTM cells (Engheta et al., 2006).

In order to retrieve the constitutive parameters of the proposed metamaterial, a unit cell
positioned between two perfect electric conductors (PEC) in x direction and two perfect
magnetic conductors (PMC) in z direction is simulated, and used to model an infinite
periodic structure (Veysi et al., 2010). The resultant scattering parameters obtained from CST
microwave studio are exerted to the Chen's algorithm (Veysi et al., 2010). The normalized
impedance (z) and refractive index (n) of the under-study medium can be calculated as
following:

$$z = \pm \sqrt{\frac{\left(1 + s_{11}\right)^2 - s_{21}^2}{\left(1 - s_{11}\right)^2 - s_{21}^2}}, \text{real}(z) \geq 0 \tag{19}$$

$$n = \frac{1}{k_0 d}\left\{\left[\left[\ln\left(e^{i n k_0 d}\right)\right]' + 2m\pi\right] - i\left[\ln\left(e^{i n k_0 d}\right)\right]''\right\} \tag{20}$$

Where

$$\text{Im}(n) \geq 0,\ e^{i n k_0 d} = \frac{S_{21}}{1 - S_{11}\dfrac{z-1}{z+1}} \tag{21}$$

$$\text{Re}(n(\omega)) = 1 + \frac{2}{\pi}\text{P.V.}\left[\int_0^\infty \frac{\omega' \text{Im}(n(\omega'))}{\omega'^2 - \omega^2}d\omega'\right] \tag{22}$$

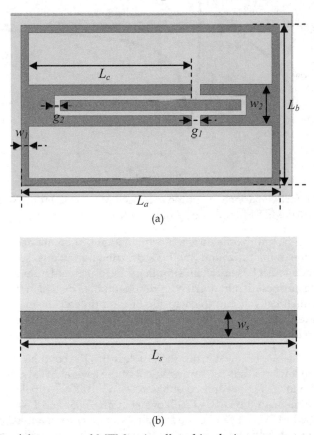

(a)

(b)

Fig. 15. Schematic of the proposed MTM unit cell and its design parameters, (a) front view, (b) back view: L_a=23.54mm, L_b=15.55mm, L_c=14.78mm, w_1=0.7mm, g_1=0.8mm, w_2=4mm, g_2= 0.5mm, w_s=2.5mm, and L_s=26.75mm. From (Rafaei et al., 2011), copyright © 2011 by the IET Microwaves, Antennas & Propagation.

The ambiguity of the value of m in (20) is resolved by using Kramers-Kronig (KK) relating the real and imaginary parts of the index of refraction (Lucarini et al., 2004). Where, $P.V.$ denotes the principal value of the integral. The effective permittivity (ε) and permeability (μ) of the medium can be expressed as: $\varepsilon=n/z$, $\mu=nz$. Fig. 16 shows the retrieved effective parameters of the proposed metamaterial cell. As can be seen, the proposed MTM cell has the permittivity that exhibits Drude behaviour at frequencies lower than 1.1GHz and Lorentz behaviour (Engheta et al., 2006) at frequencies higher than 1.1GHz.

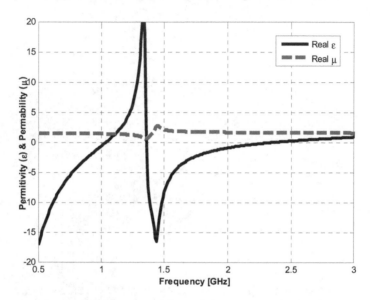

Fig. 16. Retrieved effective parameters of the Proposed MTM cell. From (Rafaei et al., 2011), copyright © 2011 by the IET Microwaves, Antennas & Propagation.

Thus, this MTM can be approximated via a combination of Lorentz and Drude models. In order to realize the miniaturization method described previous section, double-sided printed dipole antenna is chosen for its simplicity in implementation and its low profile. Fig. 17 shows the proposed miniaturized printed dipole, in which a pair of proposed MTM cells is symmetrically added to each side of the printed dipole. The proposed MTM cells and dipole are printed on a FR4 substrate with a thickness of 0.8mm and a dielectric constant of 4.4 to reduce the cost of the antenna and to make it more rigid in construction.

For the MTM cells that are far away from the dipole arms, the coupling levels of them with the dipole arms are low and thus the arrangement of the several MTM cells has no effect on the frequency behaviour of the proposed antenna. As a result, the dipole is just loaded with single cell MTM. Similar to the DNG- (Ziolkowski et al., 2003) and ENG- (Alu et al., 2003) MTMs, the proposed MTM cell can be modelled as a parallel resonant LC circuit. Thus, the proposed metamaterial cell is modelled as a resonant LC circuit parallel to the dipole, and the radiation into the free space is modelled as a resistor (Sievenpiper et al., 2006). A prototype of the proposed miniaturized dual-band printed dipole is fabricated to confirm the simulation results. Fig. 18 shows a photograph of the fabricated antenna.

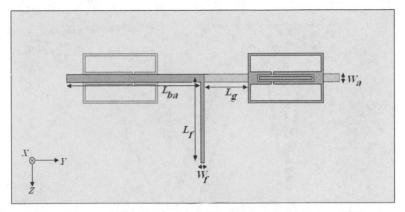

Fig. 17. Printed dipole symmetrically loaded with single cell MTM: L_{ba}=42.05mm, L_f=27.5mm, L_g=12.52mm, W_a= 2.5mm, W_f=0.8mm. From (Rafaei et al., 2011), copyright © 2011 by the IET Microwaves, Antennas & Propagation.

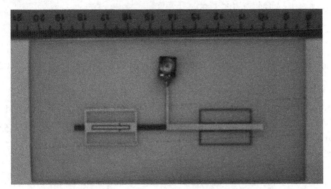

Fig. 18. Prototype of proposed miniaturized printed dipole antenna loaded with single cell MTM. From (Rafaei et al., 2011), copyright © 2011 by the IET Microwaves, Antennas & Propagation.

Fig. 19 shows the reflection coefficient of the proposed symmetrically loaded dipole with the gap length, g_1, of 0.8mm as well as the unloaded dipole antenna. As can be seen, the dipole antenna along with the loading elements provides good matching at both resonance frequencies. For comparison purposes, a simple dipole antenna loaded with lossy ENG inclusions, with the same retrieved effective parameters of the proposed MTM cell (See Fig. 16), is also simulated. As can be seen from Fig. 19, the reflection coefficient of the dipole loaded with ENG inclusions correlates nicely to that obtained for the single cell MTM loaded dipole. The co-polarized and cross-polarized radiation patterns of the proposed loaded dipole are measured at the resonant frequencies of 940MHz and 1.7GHz.

The measured and simulated radiation patterns at first and second resonant frequencies are shown in Fig. 20. As expected, the radiation patterns at both resonant frequencies are similar to that of the conventional unloaded dipole antenna. The gain of the proposed antenna at low resonant frequency is high compared to that of the other miniaturized MTM loaded dipoles (Iizuka et al., 2006; Iizuka et al., 2007; Borja et al., 2007; Rafaei et al., 2010).

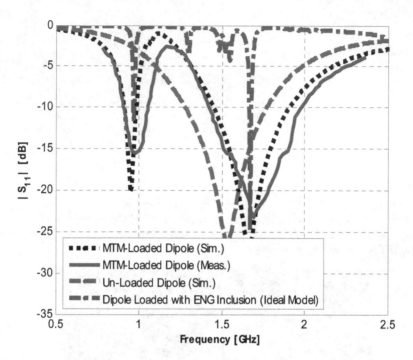

Fig. 19. Reflection coefficient of the proposed miniaturized printed dipole antenna loaded with single cell MTM. As a reference, an unloaded dipole and an ideal model of the ENG-Loaded dipole are also simulated. From (Rafaei et al., 2011), copyright © 2011 by the IET Microwaves, Antennas & Propagation.

The antenna gains at first and second resonant frequencies are -2.679dBi and 1dBi, respectively. The proposed antenna has a broad bandwidth of 15.96% at 940MHz (which is significantly wider than the bandwidth of other miniaturized MTM loaded dipoles (Iizuka et al., 2006; Iizuka et al., 2007; Borja et al., 2007; Rafaei et al., 2010)) and 32.35% at 1.7GHz. An important advantage of the proposed antenna is that the dipole length does not need to be increased to lower the resonant frequency. Consequently, a compact antenna is obtained.

Finally, the effect of the MTM location is investigated to obtain some engineering guidelines for loaded dipole designs. Thus, the loading elements move along the antenna arms and the antenna reflection coefficient is plotted in Fig. 21 for each stage. The gain, bandwidth and efficiency of the loaded dipoles with different MTM locations are also compared in Table 3.

As can be seen, the first resonant frequency remains approximately unchanged while the second one reduces as the MTM cells move away from the antenna feed point. Thus, when the MTM elements move closer to the dipole ends, the separation of the two resonances decreases. In addition, when the MTM cells are placed close to the antenna feed point, the proposed antenna cannot match very well to a 50Ω transmission line. Moreover, as can be seen from Figs. 10, 11 and 21 the single cell MTM loaded printed dipole follows closely the frequency behavior of the dipole antenna loaded with cylindrical dispersive ENG-inclusions, as d_{MIF} or L_g increases.

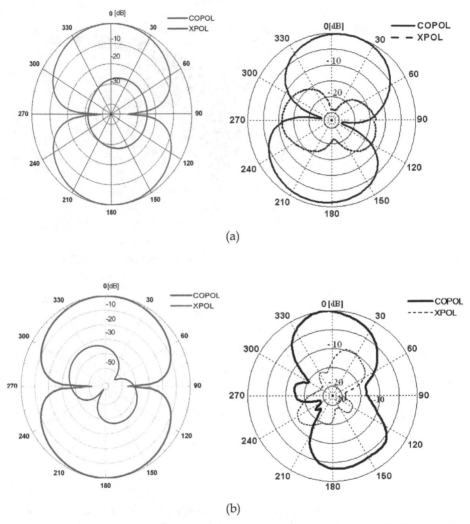

(a)

(b)

Fig. 20. Radiation patterns of the proposed printed dipole antenna at (a) 940MHz and (b) 1.7GHz. (Right hand figures are measurements). From (Rafaei et al., 2011), copyright © 2011 by the IET Microwaves, Antennas & Propagation.

In the previous design, a printed dipole was symmetrically loaded with single cell MTMs to realize a dual band operation. Here, the design parameters of the proposed dual band antenna are changed to provide a tri-band dipole antenna. As a result, two different tri-band printed dipoles are provided in the following subsections. Since changing the locations and dimensions of the MTM cells does not have any significant effect on the antenna radiation patterns, the proposed antennas radiate omnidirectional radiation patterns at all resonant frequencies. However, these are not plotted here for the sake of brevity.

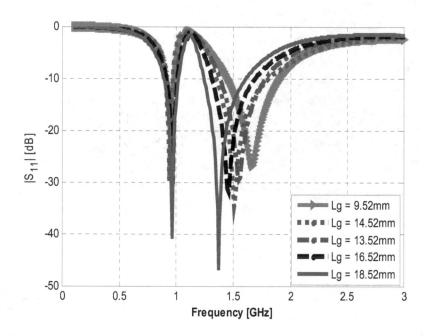

Fig. 21. The effect of MTM location on the reflection coefficient of the printed dipole antenna. From (Rafaei et al., 2011), copyright © 2011 by the IET Microwaves, Antennas & Propagation.

5. Acknowledgment

The authors would like to thank Iran Telecommunication Research Centre (ITRC) for its financial supports.

6. References

Alu A. , and Engheta N.; "Pairing an Epsilon-Negative Slab With a Mu-Negative Slab: Resonance, Tunneling and Transparency" *IEEE Trans. Antennas Propagat.*, vol. 51, pp. 2558–2571, Oct. 2003.

Balanis C. A., *Advanced Engineering Electromagnetics*. New York: Wiley, 1989.

Beggs R., Luebbers J., and Chamberlin K., "Finite difference time-domain calculation of transients in antennas with nonlinear loads," *IEEE Trans. Antennas Propag.*, vol. 41, no.5, pp.566-, May 1993.

Borja A. L., Hall P. S., Liu Q. and Iizuka H., "Omnidirectional left-handed loop antenna," *IEEE Antennas and Wireless Propagation Lett.*, Vol. 6, 495-498, 2007.

Collin R. E., and Zucker F. J., *Antenna Theory*, McGraw-Hill, 1969.

Engheta N. and Ziolkowski R., *Metamaterials: Physics and Engineering Explorations*, John Wiley & Sons Inc., 2006.

Erentok A., and Ziolkowski R. W., "Metamaterial-Inspired Efficient Electrically Small Antennas" *IEEE Trans. Antennas Propagat.*, vol. 56, pp. 691–707, March 2008.

Erentok A., Luljak P., and Ziolkowski R. W., "Antenna performance near a volumetric metamaterial realization of an artificial magnetic conductor," *IEEE Trans. Antennas Propagat.*, vol. 53, pp. 160–172, Jan. 2005.

Iizuka H. and Hall P. S., "Left-handed dipole antennas and their implementations," *IEEE Trans. Antennas Propag.*, vol. 55, no. 5, 1246–1253, May 2007.

Iizuka H., Hall P. S., and Borja A. L., "Dipole antenna with left-handed loading," *IEEE Antennas Wireless Propag. Lett.*, vol. 5, pp. 483–485, 2006.

Jafargholi A., and Kamyab M., Full-Wave Analysis of Double Positive/Double Negative Loaded Dipole Antennas, Electromagnetics, 32:2, 103-116, (2012).

Jafargholi A., and Kamyab M., *Metamaterials in Antenna Engineering, Theory and Applications*, LAP Lambert Academic Publishing, Germany, 2011.

Jafargholi A., Kamyab M., and Veysi M., "Artificial magnetic conductor loaded monopole antenna" *IEEE Antennas Wireless Propag. Lett.*, vol. 9, pp. 211-214, 2010.

Jafargholi A., Kamyab M., Rafaei M., Veysi M., "A compact dual-band printed dipole antenna loaded with CLL-based metamaterials", *International Review of Electrical Engineering*, vol. 5, no. 6, 2710-2714, 2010.

Kennedy T. F., Fasenfest K. D., Long S. A. and Williams J. T., "Modification and Control of Currents on Electrically Large Wire Structures Using Composite Dielectric Bead Elements, "IEEE Trans. Antennas Propag., vol. 54, no. 12, 3608–3613, 2006.

Kraus J. D., and Marhefka R. J., *Antennas for All Applications*, McGraw-Hill, 2002.

Liu Q., Hall P. S., and Borja A. L.," Efficiency of Electrically Small Dipole Antennas Loaded With Left-Handed Transmission Lines," *IEEE Trans. Antennas Propagat.*, vol. 57, no. 10, pp. 3009–3017, Oct. 2009.

Lucarini V., Saarinen J. J., Peiponen K. E., and Vartiainen E. M., "Kramers-Kronig Relation in Optical Materials Research", *Springer Series in optical sciences*, 2004.

Mosallaei H., and Sarabandi K., "Design and Modeling of Patch Antenna Printed on Magneto-Dielectric Embedded-Circuit Metasubstrate" *IEEE Trans. Antennas Propag.*, vol. 55, no. 1, 1031–1038, Jan 2007.

Ock J. S., and Eom H. J., "Radiation of a Hertzian Dipole in a Slotted Conducting Sphere," *IEEE Trans. Antennas Propagat.*, vol. 57, nov 12, pp. 3847–3851, Dec. 2009.

Polk C., "Resonance and Supergain Effects in Small Ferromagnetically or Dielectrically Loaded Biconical Antennas, "IRE Trans. Antennas Propag., pp. 414-423; December, 1959.

Rafaei M., Kamyab M., Jafargholi A., and Mousavi S. M., "Analytical modeling of the printed dipole antenna loaded with CRLH structures," *Progress In Electromagnetics Research B*, vol. 20, 167-186, 2010.

Rogers S. D., Butler C. M., and Martin A. Q., "Design and Realization of GA-Optimized Wire Monopole and Matching Network With 20:1 Bandwidth" *IEEE Trans. Antennas Propag.*, vol. 51, no. 3, pp. 493–502, March. 2003.

Saoudy S. A. and Hamid M., "Input admittance of a biconical antenna with wide feed gap," *IEEE Trans. Antennas Propag.*, vol. 38, no. 11, pp. 1784–1790, Nov. 1990

Shams K. M. Z. and Ali M., "*Analyses of a Dipole Antenna Loaded by a Cylindrical Shell of Double Negative* (DNG) Metamaterial," International Journal of Antennas and Propagation, vol. 2007, Article ID 97481, 10 pages.

Sievenpiper D., "Chapter11: Review of theory, fabrication, and applications of high impedance ground planes," in *Metamaterials: Physics and Engineering Explorations*, edited by Erentok A., and Ziolkowski R. W.

3

Designs of True-Time-Delay Lines and Phase Shifters Based on CRLH TL Unit Cells

J. Zhang, S.W. Cheung and T.I. Yuk
Department of Electrical and Electronic Engineering,
The University of Hong Kong, Hong Kong,
China

1. Introduction

1.1 Brief description on definition of electromagnetic metamaterials

Electromagnetic (EM) Metamaterials have specific EM properties that cannot be found in nature (Caloz & Itoh, 2006). These specific EM properties are obtained by the artificial structures, rather than the composition of the metamaterials, and affect the propagations of EM waves when the average structure sizes are much smaller than the guided wavelengths λ_g, i.e., at least smaller than $\lambda_g/4$. Since the propagation of an EM wave is related to its electric and magnetic fields, the EM properties of Electromagnetic (EM) Metamaterials can be described by using its permittivity ε and permeability μ.

Fig. 1 shows the four possible combinations of permittivity and permeability of materials. In quadrant I, we can find materials such as isotropic dielectrics which have $\varepsilon > 0$ and $\mu > 0$. In quadrant II, where $\varepsilon < 0$ and $\mu > 0$, we can find materials like plasmas to have such properties. While in quadrant IV, we also can find ferromagnetic materials to have $\varepsilon > 0$ and $\mu < 0$. However, in quadrant III where both permittivity and permeability are negative, i.e. $\varepsilon < 0$ and $\mu < 0$, there is no natural material having such properties. However, one can construct artificial structures that have negative permittivity and permeability, and these artificial materials are called left-handed (LH) materials as will be explained in the following sections.

1.2 Left-handed (LH) materials

When an EM wave propagates in a conventional material, the electric field **E** and magnetic field **H** are orthogonal to each other and also orthogonal to the wave-vector **k** which has the same direction as the power flow density (also known as Poynting vector **S**). These three vectors, **E**, **H** and **k**, form a right-hand (RH) triplet, so conventional materials can also be called right-handed (RH) materials. In the 1960s, Russian physicist Viktor Veselago theoretically investigated the existence of materials with negative permittivity ε and also negative permeability μ (Veselago, 1968). He speculated that such materials would also satisfy Maxwell's equations but allow the electric field **E**, magnetic field **H** and wave-vector **k** of an EM wave to form a left-handed (LH) triplet. For this reason, the term 'left-handed' was used to describe these theoretical materials. Based on the fact that, in these materials, **E**,

H and **k** formed a LH triplet and **E**, **H** and **S** formed a RH triplet, Veselago showed that, for a uniform plane wave propagating in such materials, the direction of the wave vector **k** would be anti-parallel to the direction of the Poynting vector **S**. The phenomenon is in contrary to the case of plane wave propagation in conventional or RH materials.

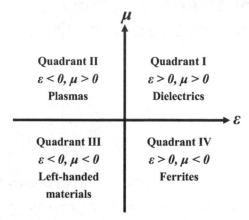

Fig. 1. Permittivity-permeability diagram

1.3 Wave propagation in LH materials by Maxwell's equations

The general form of time-varying Maxwell's equations can be written as (Ulaby, 2004):

$$\nabla \times E = -\frac{\partial B}{\partial t} \tag{1a}$$

$$\nabla \times H = J + \frac{\partial D}{\partial t} \tag{1b}$$

$$\nabla \bullet D = \rho \tag{1c}$$

$$\nabla \bullet B = 0 \tag{1d}$$

where ∇ : vector operator $\frac{\partial}{\partial x}\hat{x} + \frac{\partial}{\partial y}\hat{y} + \frac{\partial}{\partial z}\hat{z}$,

- $\bullet$: vector dot product,
- **E:** electric field intensity, in V/m,
- **H:** magnetic field intensity, in A/m,
- **D:** electric flux density, in C/m²,
- **B:** magnetic flux density, in W/m²,
- **J:** electric current density, in A/m², and
- ρ : electric charge density, in C/m³.

The relationships between the field intensities **E** and **B**, flux densities **D** and **H** and current density **J** are (Ulaby, 2004):

$$D = \varepsilon E, \quad B = \mu H, \quad J = \sigma E \tag{2}$$

where ε, μ and σ are the electric permittivity (also called the dielectric constant), permeability and conductivity, respectively, of the material under consideration.

Consider a wave having a single frequency ω. Then by introducing the time factor $e^{+j\omega t}$ to (1), the time derivatives in Maxwell's equations of (1) can be replaced by $j\omega$ and Maxwell's equations can be re-written as:

$$\nabla \times E = -j\omega\mu H \tag{3a}$$

$$\nabla \times H = (\sigma + j\omega\varepsilon)E \tag{3b}$$

$$\nabla \bullet E = \rho / \varepsilon \tag{3c}$$

$$\nabla \bullet H = 0 \tag{3d}$$

In free space which is lossless, J, σ and ρ are all zero and (3) becomes:

$$\nabla \times E = -j\omega\mu H \tag{4a}$$

$$\nabla \times H = j\omega\varepsilon E \tag{4b}$$

$$\nabla \bullet E = 0\rho / \varepsilon \tag{4c}$$

$$\nabla \bullet H = 0 \tag{4d}$$

It can be readily shown from (4) that

$$\nabla^2 E + k^2 E = 0 \tag{5a}$$

$$\nabla^2 H + k^2 H = 0 \tag{5a}$$

where $k = 2\pi/\lambda$ is the wave number (which is real if ε and μ are real) with λ being the wavelength.

(5) is known as the wave equation which has many solutions. A plane wave is a wave having a constant phase over a set of planes, while a uniform-plane wave is a wave having both magnitude and phase constant. An EM wave in free space is a uniform-plane wave having the electric field **E** and magnetic field **B** mutually perpendicular to each other and also to the direction of propagation, i.e., a transverse electromagnetic (TEM) wave. For convenience and without lost of generality, assume that the EM wave considered is propagating in the +z direction in free space. Under these conditions, the solution for (5) is (Ulaby, 2004):

$$E(z) = E_0 e^{-j k \bullet z} \tag{6a}$$

$$H(z) = e_k \times \frac{1}{\eta_0} E(z) = \frac{e_k \times E_0(z)}{\eta_0} e^{-j k \bullet z} \tag{6b}$$

where $E_0 = E_m e^{j\phi} e_x$ (with $E_m = |E_0|$ and e_x being the unit vector along +x direction)
- η_0 : free space impendence 377 Ω
- **k**: wave vector having magnitude of k at **+z** direction .

Substituting (6a) and (6b) into (4a) and (4b), respectively, yields

$$k \times E = \omega \mu H \tag{7a}$$

$$k \times H = -\omega \varepsilon E \tag{7b}$$

in which, for positive values of ε and μ, the triplet (**E**, **H**, **k**) can be used to produce an orthogonality diagram shown in Fig. 2(a) using our right hand (RH). However, if ε and μ are both negative, (7) becomes:

$$k \times E = -\omega |\mu| H \tag{8a}$$

$$k \times H = \omega |\varepsilon| E \tag{8b}$$

in which now the triplet (**E**, **H**, **k**) forms an orthogonality diagram shown in Fig. 2(b) using our left hand (LH). This result verifies Velago's speculation and this is the primary reason for materials with negative ε and μ to be called LH materials.

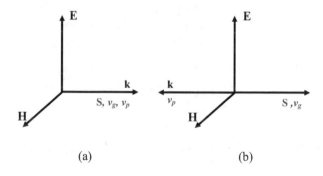

(a)   (b)

Fig. 2. Orthogonality diagram of E, H, k for uniform plane wave in (a) RH material and (b) LH material

The power flow density of an EM wave, also known as the Poynting vector, is defined as (Ulaby, 2004):

$$S = E \times H \tag{9}$$

(9) indicates that the power flow density is only determined by **E** and **H** but not the signs of ε and μ. Thus in both RH and LH materials, the triplets (**E**, **H, S**) have the same form of orthogonality, i.e., following our right hand, as shown in Figs. 2(a) and 2(b), and the directions of energy flow are the same.

The group velocity v_g of an EM wave is given by (Ulaby, 2004):

$$v_g = \frac{\partial \omega}{\partial k} \tag{10}$$

which expresses the velocity of power flow and so has the same direction as the Poynting vector **S**, as shown in Fig. 2.

The phase velocity v_p is the velocity of wave-front given by (Ulaby, 2004):

$$v_p = \frac{\omega}{k} \tag{11}$$

and so has the same sign as k, i.e., $v_p > 0$ for $k > 0$ in RH materials and $v_p < 0$ for $k < 0$ in LH materials. As a result, the v_p in LH materials and in RH materials are anti-parallel as indicated in Fig. 2.

Table 1 summarizes the aforementioned analysis for a plane uniform EM wave in the RH and LH materials.

	triplet (**E,H,k**)	k	v_p	triplet (**E,H,S**)	v_g
RH material	RH orthogonality	> 0	> 0	RH orthogonality	> 0
LH material	LH orthogonality	< 0	< 0	RH orthogonality	> 0

Table 1. Characteristics of RH and LH material

2. Transmission line (TL) approach of metamaterials

2.1 Left-handed transmission lines (LH TLs)

A microstrip transmission line fabricated on a substrate is shown in Fig. 3(a). The transmission line is also called a right-handed transmission line (RH TL) because when an EM wave travelling through it, the triplet (**E, H, k**) forms a RH orthogonality, as will be shown later. The equivalent circuit of the transmission line is shown in Fig. 3(b), where L_R models the RH series inductance along the transmission line and C_R models the RH shunt capacitance between the transmission line and the ground on the other side of the substrate (Pozar, 2004).

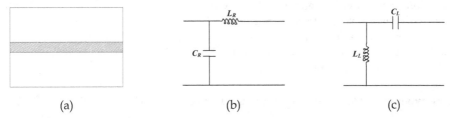

| (a) | (b) | (c) |

Fig. 3. (a) Structure of RH TL (b) equivalent circuit of RH TL and (c) equivalent circuit of LH TL

To simplify our analysis, we assume that the transmission line is lossless. The complex propagation constant γ, the phase constant β, the phase velocity v_p and the group velocity v_g of the RH TL are, respectively, given by (Pozar, 2004):

$$\gamma = j\beta = \sqrt{ZY} = j\omega\sqrt{L_R C_R} \tag{12a}$$

$$\beta = \omega\sqrt{L_R C_R} > 0 \tag{12b}$$

$$v_p = \frac{\omega}{\beta} = \frac{1}{\sqrt{L_R C_R}} > 0 \tag{12c}$$

$$v_g = \frac{\partial \omega}{\partial \beta} = \frac{1}{\sqrt{L_R C_R}} > 0 \tag{12d}$$

By making a duality of the equivalent circuit in Fig. 3(b), i.e. replacing the series inductance L_R with a series capacitance C_L and the shunt capacitance C_R with a shunt inductance L_L, we can have the equivalent circuit of a left-handed transmission line (LH TL) (Lai et al., 2004) shown in Fig. 3(c). It is called a LH TL simply because when an EM wave travelling through it, the triplet (**E**, **H**, **k**) forms a LH orthogonality, as will be shown later. Now the complex propagation constant γ, the phase constant β, the phase velocity v_p and the group velocity v_g of the LH TL can be found as

$$\gamma = j\beta = \sqrt{ZY} = -j\omega\frac{1}{\sqrt{L_L C_L}} \tag{13a}$$

$$\beta = -\frac{1}{\omega\sqrt{L_L C_L}} < 0 \tag{13b}$$

$$v_p = \frac{\omega}{\beta} = -\omega^2\sqrt{L_L C_L} < 0 \tag{13c}$$

$$v_g = \frac{\partial \omega}{\partial \beta} = \omega^2\sqrt{L_L C_L} > 0 \tag{13d}$$

Here the phase constant β (= $2\pi/\lambda$) in (13b) (equivalent to the wave number k used in wave propagation previously) is negative and so the phase velocity v_p in (13c) associated with the direction of phase propagation is negative. However, the group velocity v_g in (13d) indicating the direction of power flow (Poynting vector S) remains positive. This characteristic agrees with that of LH materials shown in Table 1, so the LC circuit shown in Fig. 3(c) can be used to realize LH materials.

2.2 Composite right/left-handed transmission line (CRLH TL)

The transmission line structure shown in Fig. 4(a) was proposed in (Lai et al., 2004) to realize the series capacitance C_L and shunt inductance L_L in Fig. 3(c) for the LH materials (Lai et al., 2004). The structure consists of a series inter-digital capacitors to realize the series capacitance C_L and a via shunted to ground on the other of the substrate to realize the shunt inductance L_L. However, when an EM wave travels along the structure, the current flowing along the upper metal trace induces a magnetic field, creating an inductive effect. This effect is modeled by the series inductance L_R in Fig. 4(b). Moreover, the potential difference generated between the upper metal trace and the ground plane on the other side produces an electric field, creating a capacitive effect. This effect is modeled by the shunt capacitance C_R in Fig. 4(b). Since the inductive and capacitive effects caused by the series inductance L_R and shunt capacitance C_R, respectively, cannot be avoided in practical implementation of LH TLs, the term "composite right/left-handed transmission line" (CRLH TL) is used to describe such a structure.

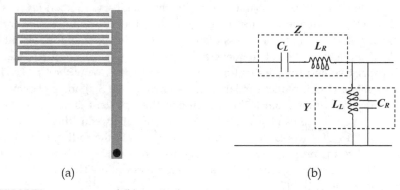

(a)   (b)

Fig. 4. (a) CRLH TL structure and (b) equivalent circuit

2.3 Dispersion diagrams

Here, we show that the structure of Fig. 4(a) indeed has LH and RH properties shown in Table 1. The complex propagation constant of the CRLH TL structure in Fig. 4(b) can be written as (Caloz & Itoh, 2006; Lai et al., 2004)

$$\gamma = \alpha + j\beta = \sqrt{ZY} \tag{14}$$

where α and are β are the attenuation and phase constants, respectively. Assume the structure, as represented by Fig. 4(a), is lossless, and so has no attenuation, i.e. $\alpha = 0$ in (14). The propagation constant is an imaginary number:

$$\gamma = j\beta = \sqrt{ZY} = \sqrt{(j\omega L_R + \frac{1}{j\omega C_L})(j\omega C_R + \frac{1}{j\omega L_L})} \tag{15a}$$

$$= \sqrt{(-1)}\sqrt{\omega^2 L_R C_R + \frac{1}{\omega^2 L_L C_L} - (\frac{L_R}{L_L} + \frac{C_R}{C_L})} \tag{15b}$$

$$= js(\omega)\sqrt{\omega^2 L_R C_R + \frac{1}{\omega^2 L_L C_L} - (\frac{L_R}{L_L} + \frac{C_R}{C_L})} \tag{15c}$$

where

$$s(\omega) = \pm 1 \tag{16}$$

In the equivalent circuit of Fig. 4(b), the components C_L and L_R form a series-tune circuit with a resonant frequency at $1/\sqrt{L_R C_L}$. At frequencies larger than $1/\sqrt{L_R C_L}$, the circuit is inductive. While the components C_R and L_L form a parallel-tune circuit resonating at $1/\sqrt{L_L C_R}$. At frequencies larger than $1/\sqrt{L_L C_R}$, the circuit is capacitive. Thus, at frequencies $\omega > \max(1/\sqrt{L_R C_L}, 1/\sqrt{L_L C_R})$, the series-tune circuit is inductive and the parallel-tune circuit is capacitive. The CRLH TL structure will have an equivalent-circuit model similar to one shown in Fig. 3(b) and so behaves like a RH TL. Under this condition, from (12b), the propagation constant is a positively imaginary number, i.e., $\gamma = j\beta$, and so $s(\omega) = +1$ in (16). At frequencies less than $1/\sqrt{L_R C_L}$, the series-tune circuit formed by C_L and L_R is capacitive. At frequencies less than $1/\sqrt{L_L C_R}$, the parallel-tune circuit formed by C_R and L_L is inductive. Thus at frequencies $\omega < \min(1/\sqrt{L_R C_L}, 1/\sqrt{L_L C_R})$, the series-tune circuit is capacitive and the parallel-tune circuit is inductive. Now the CRLH TL structure has an equivalent-circuit model similar to one shown in Fig. 3(c) and so behaves like a LH TL. Under this condition, from (13b), the propagation constant is a negatively imaginary number i.e., $\gamma = -j\beta$ and so $s(\omega) = -1$ in (16). Between these two limits, i.e., $\min(1/\sqrt{L_R C_L}, 1/\sqrt{L_L C_R}) < \omega < \max(1/\sqrt{L_R C_L}, 1/\sqrt{L_L C_R})$, the radicand in (15c) is purely imaginary and so the propagation constant is purely real at $\gamma = \beta$. There is only attenuation in the CRLH TL structure which behaves as a stop-band filter. This stop band is a unique characteristic of the CRLH TL.

Therefore, the propagation constant of a CRLH TL structure can be re-written from (15) as:

$$
\gamma(\omega) = \begin{cases}
j\beta(\omega) = j\sqrt{\omega^2 L_R C_R + \dfrac{1}{\omega^2 L_L C_L} - \left(\dfrac{L_R}{L_L} + \dfrac{C_R}{C_L}\right)} \\
\quad \text{If } \omega > \max(\dfrac{1}{\sqrt{L_R C_L}}, \dfrac{1}{\sqrt{L_L C_R}}) \\[6pt]
j\beta(\omega) = -j\sqrt{\omega^2 L_R C_R + \dfrac{1}{\omega^2 L_L C_L} - \left(\dfrac{L_R}{L_L} + \dfrac{C_R}{C_L}\right)} \\
\quad \text{If } \omega < \min(\dfrac{1}{\sqrt{L_R C_L}}, \dfrac{1}{\sqrt{L_L C_R}}) \\[6pt]
\beta(\omega) = \sqrt{-\omega^2 L_R C_R - \dfrac{1}{\omega^2 L_L C_L} + \left(\dfrac{L_R}{L_L} + \dfrac{C_R}{C_L}\right)} \\
\quad \text{If } \min(\dfrac{1}{\sqrt{L_R C_L}}, \dfrac{1}{\sqrt{L_L C_R}}) < \omega < \max(\dfrac{1}{\sqrt{L_R C_L}}, \dfrac{1}{\sqrt{L_L C_R}})
\end{cases}
\tag{17}
$$

The dispersion diagrams of the RH TL, LH TL, and CRLH TL plotted using (12b), (13b) and (17), respectively, are shown in Fig. 5 (Caloz & Itoh, 2006; Lai et al., 2004). For the RH TL, Fig. 5(a) shows that the group velocity (i.e., $v_g = d\omega/d\beta$) is positive and has values only for $\beta > 0$. Since phase velocity is defined as $v_p = \omega/\beta$, Fig. 5(a) shows that the RH TL has $v_p > 0$. Fig. 5(b) shows that the LH TL also has a positive group velocity ($v_g = d\omega/d\beta$) but only for $\beta < 0$ which leads to $v_p < 0$ (because $v_p = \omega/\beta$). These characteristics agree with those of the RH and LH materials shown in Table 1. Thus the circuit models in Figs. 3(b) and 3(c) realized using transmission lines have the EM properties of the RH and LH materials, respectively. In Fig. 5(c), the regions for $\beta > 0$ and $\beta < 0$ are known here as the RH and LH regions, respectively.

Now, consider the dispersion diagram in Fig. 5(c). It can be seen that the CRLH TL behaves like a LH TL for $\beta < 0$ ($v_p < 0$, $v_g > 0$) and like a RH TL for $\beta > 0$ ($v_p > 0$, $v_g > 0$). Thus the circuit of Fig. 4(b) indeed has both the LH and RH properties. Fig. 5(c) shows that the CRLH TL has quite small group velocities in the LH region and much large group velocities in the RH region. The LH region with low group velocities can be used to implement TTDLs with high time-delay efficiencies as described later. Moreover, the CRLH TL has a stopband in the frequency range: $\min(1/\sqrt{L_R C_L}, 1/\sqrt{L_L C_R}) < \omega < \max(1/\sqrt{L_R C_L}, 1/\sqrt{L_L C_R})$, where $\beta = 0$, leading to a zero group velocity v_g and meaning zero power flow.

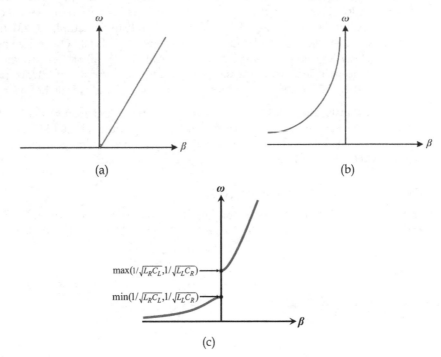

Fig. 5. Dispersion diagrams of (a) RH TL, (b) LH TL and (c) CRLH TL

3. True-time-delay lines (TTDLs) using CRLH TL unit cells

3.1 Introduction

True-time-delay lines (TTDLs) are widely employed in various microwave devices and subsystems. They find applications in phased arrays, feed-forward amplifiers, delay-lock loops, phase noise measurement systems and oscillators (Lee, 2004). There are different approaches to implement the TTDLs. For example, in magnetostatic wave (MSW) TTDLs (Fetisov & Kabos, 1998) and surface-acoustic wave (SAW) TTDLs (Smith & Gerard, 1969), the group velocities were slowed down by transferring the microwave signals into the magnetostatic waves and surface acoustic waves, respectively. These designs have quite high time-delay efficiencies, but the bulky and complicated transducers required are not conducive for planar microwave circuits. Moreover, SAW TTDLs have extremely narrow bandwidths of only several MHz and MSW TTDLs have very large insertion losses. Optical TTDLs (Xu et al., 2004) have very small insertion losses, so we can use very long optical fiber cables to achieve very large TDs. However, optical TTDLs also need complicated transducers to transfer the microwave signals into the optical waves.

Microstrip lines, with the advantages of simplicity in feeding and compatibility with planar circuits, are widely used in communications systems, particularly for small communications devices. A simple RH TL can also be used to implement TTDL. However, due to the low time-delay efficiency, it is difficult to achieve a long TD using RH TL. To increase the time-delay efficiency of RH TL, various slow-wave structures have been proposed. Some employ periodic discontinuities such as the Electromagnetic Bandgap Structures (EBG) (Kim & Drayton, 2007) and defected ground structure (DGS) (Woo et al., 2008). Others use periodic equivalent LC networks (Zhang & Yang, 2008; Zhang et al., 2011). All of these designs adapt the same basic concept,of increasing the series inductance and shunt capacitance per unit length, hence the effective dielectric, to reduce the guided wavelength of the EM waves.

As described previously, the CRLH TL has a unique dispersion characteristic, i.e., having small group velocities in the LH region. Thus a CRLH TL operating in the LH region can be used to realize a TTDL with a high time-delay efficiency. In this section, a TTDL using four symmetrical CRLH TL unit cells is studied.

3.2 TTDL realized by transmission line (TL)

The true-time delay (TTD) of a transmission line (TL) is the time it takes for an EM wave to travel through it, so a TL can be used to design a TTDL. For a TL with a length of L, the TTD is:

$$\tau = L / v_g \tag{18}$$

where v_g is the group velocity given by

$$v_g = d\omega / d\beta \tag{19}$$

with β and ω being the propagation constant and frequency in rad/s, respectively. (18) and (19), show that the TTD is inversely proportional to the group velocity v_g. For a given length of TL, the smaller is the v_g, the longer will be the TD.

3.3 Symmetrical CRLH TL unit cell

To design a TTDL with a large TTD, we propose to cascade a number of CRLH TL structures together, each having a small group velocity. For matching purpose, the CRLH TL structure has to be designed so that $S_{11} = S_{22}$ over the operating frequency band. However, this is not easy to do using the CRLH TL structure shown in Fig. 4(a) because adjusting any of the structural parameters will change both S_{11} and S_{22}. To overcome such difficulty, we propose a new structure as shown in Fig. 6 which has a symmetrical structure and is here called a symmetrical CRLH TL unit cell (Zhang et al., 2009). Compared with the CRLH TL structure as shown in 4(a), our proposed CRLH TL unit cell has two stubs, instead of one stub, having a grounded via, making the whole structure centrosymmetrical. To deign such unit cell for use in our TTDL, we only need to match the input impedance to a 50-Ω coaxial cable, which is relatively easy to do. Once this matching is designed, due to its symmetrical structure, the two Z-parameters, Z_{11} and Z_{22}, will be the same and equal to 50 Ω. The symmetry of the unit cell leads to an equivalent-symmetrical π-model shown in Fig. 6(c) in which L_R models the RH series inductance along all the horizontal fingers, C_L models the LH coupling capacitance between the fingers, $2L_L$ models the LH shunt inductance of each stub having a via at the end to the ground, and $C_R/2$ models the RH shunt capacitance between the fingers and the ground on each side of the unit cell.

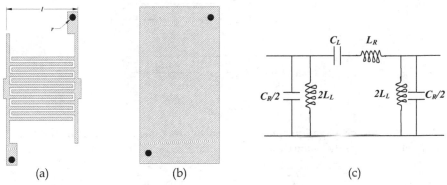

(a)   (b)   (c)

Fig. 6. Proposed CRLH TL unit cell. (a) Top view, (b) bottom, and (c) equivalent circuit.

Here, we will show that the CRLH TL unit cell in Fig. 6(a) has the same dispersion diagram shown in Fig. 5 (c), so that we can have the LH region to operate the unit cell. The expressions for L_R, C_R, L_L and C_L in Fig. 6(c) are, respectively, given by (Bahl, 2001; Marc & Robert, 1991; Pozar, 2004):

$$L_R = \frac{Z_0\sqrt{\varepsilon_{re}}}{Nc}l \tag{20a}$$

$$C_R = \frac{N\sqrt{\varepsilon_{re}}}{Z_0 c}l \tag{20b}$$

$$C_L = (\varepsilon_r + 1)l\left\{4.409(N-3)\tanh\left[0.55\left(\frac{h}{W}\right)^{0.45}\right] + 9.92\tanh\left[0.52\left(\frac{h}{W}\right)^{0.5}\right]\right\} \times 10^{-12} \tag{20c}$$

$$L_L = \frac{\mu_0}{2\pi}\left[h \cdot \ln\left(\frac{h + \sqrt{r^2 + h^2}}{r}\right) + \frac{3}{2}(r - \sqrt{r^2 + h^2})\right] + \frac{Z_0\sqrt{\varepsilon_{re}}}{c}l' \tag{20d}$$

where h is the thickness of the substrate, r is the radius of the ground via, ε_r is the relative dielectric constant, ε_{re} is the effective dielectric constant, l' is the distance from the ground via to the port, l is the length of the finger, N is the number of fingers, W is the width of all the fingers together and Z_0 is the characteristic impedance of each of the fingers. To derive the complex propagation constant of this symmetrical CRLH TL unit cell, we separate the L_R and C_L in Fig. 6(c) into two capacitances $2C_L$ and two inductances $L_R/2$, respectively. This results in two sub-circuits as shown in Fig. 7, each having the same propagation constant γ'. The total propagation constant γ is the sum of the propagation constants for these two sub-circuits, i.e., $\gamma = 2\gamma'$. In Fig. 7, the propagation constant of the symmetrical CRLH TL unit cell is:

$$\gamma = j\beta = 2\gamma' = 2\sqrt{ZY} = 2\sqrt{(j\omega\frac{L_R}{2} + \frac{1}{j\omega2C_L})(j\omega\frac{C_R}{2} + \frac{1}{j\omega2L_L})}$$

$$= \sqrt{(-1)}\sqrt{\omega^2 L_R C_R + \frac{1}{\omega^2 L_L C_L} - (\frac{L_R}{L_L} + \frac{C_R}{C_L})}$$

$$= js(\omega)\sqrt{\omega^2 L_R C_R + \frac{1}{\omega^2 L_L C_L} - (\frac{L_R}{L_L} + \frac{C_R}{C_L})} \tag{21}$$

where $s(\omega) = \pm1$

(21) shows that the propagation constant is exactly the same as (15c). Thus the symmetrical CRLH TL unit cell has the same dispersion diagram shown in Fig. 5(c).

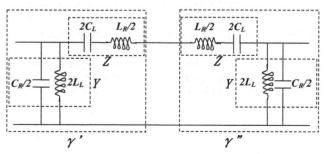

Fig. 7. Equivalent circuit of Fig. 6(c)

3.4 TTDL realization using symmetrical CRLH TL unit cells

From (18) and (19), the TD of a TL is given by

$$\tau = \frac{L}{v_g} = \frac{L}{d\omega/d\beta} = \frac{Ld\beta}{d\omega} = \frac{Ang(s_{21})}{2\pi df} \tag{22}$$

where $Ang(S_{21})$ is the phase of S_{21}. Figure 5(c) shows that the CRLH TL has quite small group velocities v_g in the LH region and much higher group velocities in the RH region. Thus the LH region with low group velocities can be used to implement TTDLs with longer TTDs and high time-delay efficiencies as indicated in (22). It can be see that (22) also indicate that, for a given length L, the larger is the slope of phase response, i.e., $Ang(S_{21})/df$, the longer will be the TD or the higher will be the time-delay efficiency. Moreover, by adjusting

L_R, C_R, L_L and C_L using the structural parameters through (20), the CRLH TL unit cell can also be designed to operate at different frequencies and TDs.

3.5 Simulation and measurement results

3.5.1 Single CRLH TL unit cell

The proposed symmetrical CRLH TL unit cell shown in Fig. 6 has been designed with a center frequency at around 3 GHz on a Rogers substrate, RO4350, with a thickness of 0.762 and a permittivity of 3.48 using computer simulation. It has a total a length of 6.8 mm. The design is optimized for the criteria for wide impedance bandwidth, small insertion and large phase response. The simulated results on return loss, $-10\log|S_{11}|$, and insertion loss, $-10\log|S_{21}|$, of the CRLH TL unit cell are shown in Fig. 8(a), while the phase response, $(\mathrm{Ang}(S_{21}))$, shown in Fig. 8(b). For comparison, the results of a RH TL with the same length of 6.8 mm are also shown in the same figure. Figure 8(a) shows that the CRLH unit cell has an operating bandwidth of 2.2-4.0 GHz, a return loss of large than 15 dB and insertion loss of less than 1 dB within the operating bandwidth. The slope of phase response, as shown in Fig. 8(b), is 41.3 degree/GHz, about 3 times larger that of the RH TL at 13.3 degree/GHz. Thus we can expect that the CRLH TL unit cell can achieve a TTD 3.1 times longer than that of a RH TL for the same length and so has the time delay efficiency 3.1 times higher than that of the RH TL.

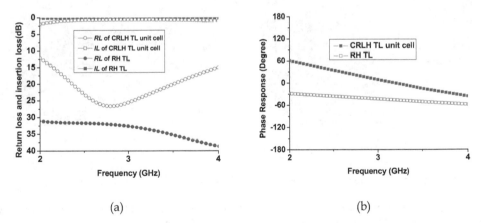

(a)   (b)

Fig. 8. Simulated (a) return loss (RL) and insertion loss (IL) and (b) phase response of CRLH TL unit cell and RH TL

3.5.2 Multi-CRLH TL unit cells

A TTDL constructed by cascading four symmetrical CRLH unit cells, with a total length of 30 mm, is shown in Fig. 9(a). The TTDL has been designed, studied and optimized using computer simulation. The final design has also been implemented on a Rogers substrate, RO4350, with a thickness of 0.762 and a permittivity of 3.48 and measured for verification of simulation results. For comparison, a TTDL implemented using a RH TL with the same length of 30 mm, as shown in Fig. 9(b), has also been designed and simulated using the same substrate. The prototype-modules of the two TTDLs with the same dimension of 30 mm×15 mm×2 mm are shown in Fig. 10.

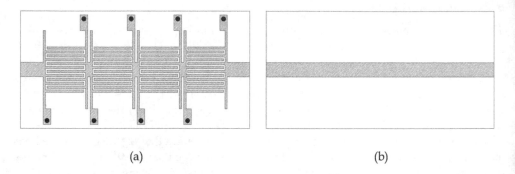

(a)   (b)

Fig. 9. TTDLs using (a) four CRLH TL unit cells and (b) RH TL

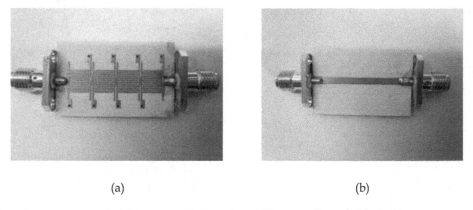

(a)   (b)

Fig. 10. Prototypes of TTDLs using (a) four CRLH TL unit cells and (b) RH TL

The simulated and measured return losses, insertion losses and TTDs of the two TTDLs are shown in Fig. 11. It can be seen that the simulated and measured results show good agreements. Figure 11(a) show that the measured return losses and the insertion losses of the TTDLs are more than 15 dB and less than 1 dB, respectively, across the frequency band from 2.2 – 3.7 GHz. The measurement results in Fig. 11(b) show that the TTDL using CRLH TL unit cells achieves a TTD of 510 ps, about 3.2 times larger than the RH TL having a TTD of 160 ps. This result is consistent with the result obtained for a single CRLH TL unit cell. The measurement results in Fig. 11(b) show that the maximum TTD error for the TTDL using the CRLH TL unit cells is about -21 ps or -4.1% at the frequency of 2.4 GHz. For the TTDL using RH TL, although the maximum TD error is about -7.8 ps at the frequency of 2.8 GHz, the percentage is higher at -4.8%.

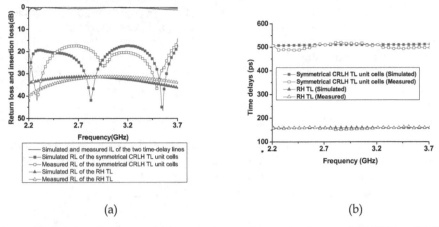

(a) (b)

Fig. 11. Simulated and measured (a) return loss and insertion loss and (b) TTDs of TTDLs using CRLH TL unit cells and RH TL

4. Digital-phase shifters using CRLH unit cells

4.1 Introduction

Phase shifters are essential components in radar and phased array systems. In the designs of phase shifters, insertion loss, size and power-handling capability are major factors for considerations. Phase shifters can be classified into passive and active. Passive phase shifters can be implemented using ferrite technology (Adam et al., 2002) to achieve higher power capabilities, but they have large sizes and heavy weights. Active phase shifters implemented using solid-state devices such as FET and CMOS technologies have smaller sizes. However, their low power-handling capabilities and nonreciprocal characteristics limit their applications.

In digital-phase shifters, phase shift is usually obtained by switching between two transmission lines of different lengths or between lumped-element low-pass and high-pass filters (Keul & Bhat, 1991). Usually, the switches are implemented using solid-state devices such as PIN diodes which have typical power-handling capabilities of just a few Watts and this limits the power-handling capabilities of the phase shifters. Moreover, if an n-bit phase shifter is constructed by cascading several phase shifters together to provide the required phase shifts, the physical size and insertion loss will be undoubtedly increased.

Recently, different design approaches of phase shifters based on using metamaterials have been proposed and studied (Antoniades & Eleftheriades, 2003; Damm et al., 2006; Kim et al., 2005; Kholodnyak et al., 2006; Lapine et al., 2006; Vendik et al., 2009;). These designs share one of the major drawbacks, i.e., the power-handling capability is limited by the power-handling capabilities of the switches, tunable diodes and tunable capacitors used in the designs. A phase shifter based on CRLH TL employing MEMS has also been proposed and studied (Monti et al., 2009), but the reliability of the design needs more studies (Rebeiz et al., 2002).

In this section, we present an approach for the design of n-bit phase shifters using the CRLH TL unit cells. The phase shifters designed using this approach have the advantages of compact size, high power-handling capability, low insertion loss, arbitrary phase-shift range and arbitrary step size. PIN diodes mounted on the fingers of the CRLH TL unit cell are used as switches to control the phase shift. Different phase shifts are achieved by using different states of these switches and the controlling bits are used to select one of those switch states for the required phase shift.

4.2 Description of CRLH TL unit cell using ABCD-parameters

The symmetrical CRLH TL unit cell used for the designs of digital-phase shifters is shown in Fig. 12(a) with the equivalent π-model circuit shown in Fig. 12(b). The symmetrical CRLH TL unit cell and the equivalent π-model circuit are similar to those shown in Fig. 6(a), thus the expressions for L_R, C_R, L_L and C_L are same as given in (20a)-(20d).

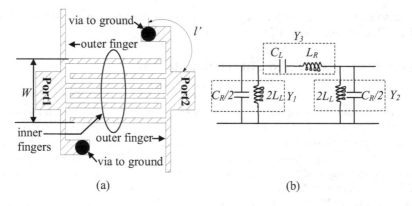

(a)   (b)

Fig. 12. Structure and equivalent circuit of symmetrical CRLH TL unit cell.

The transmission ABCD matrix of the π-model circuit in Fig. 12 (b) is (M.D. Pozar, 2004):

$$\begin{bmatrix} A & B \\ C & D \end{bmatrix} = \begin{bmatrix} 1 + Y_2 / Y_3 & 1 / Y_3 \\ Y_1 + Y_2 + Y_1 Y_2 / Y_3 & 1 + Y_1 / Y_3 \end{bmatrix}$$ (23)

where $Y_1 = Y_2 = j\omega C_R / 2 + 1 / 2 j\omega L_L$ and $Y_3 = \dfrac{1}{j\omega L_R + 1 / j\omega C_L}$.

Using (23), the expressions for S_{11}, S_{21} and phase incursion Ang(S_{21}) can be easily obtained, respectively, as:

$$S_{11} = \frac{(\frac{b}{Z_0})^2 - Z_0^2 a^2 [2 - ab]^2}{4[1 - ab]^2 + [Z_0 a^2 b - 2Z_0 a - (\frac{b}{Z_0})]^2} + j \frac{2[1 - ab][2Z_0 a - \frac{b}{Z_0} - a^2 b Z_0]}{4[1 - ab]^2 + [Z_0 a^2 b - 2Z_0 a - \frac{b}{Z_0}]^2}$$ (24a)

$$S_{21} = \frac{4[1-ab]}{4[1-ab]^2 + [Z_0a^2b - 2Z_0a - \frac{b}{Z_0}]^2} - 2j\frac{[Z_0a^2b - 2Z_0a - \frac{b}{Z_0}]}{4[1-ab]^2 + [Z_0a^2b - 2Z_0a - \frac{b}{Z_0}]^2} \tag{24b}$$

and

$$Ang(S_{21}) = \tan^{-1}[-\frac{Z_0a^2b - 2Z_0a - \frac{b}{Z_0}}{2(1-ab)}] \tag{24c}$$

$$\text{where } a = \frac{1-\omega^2 L_L C_R}{2\omega L_L} \text{ and } b = \frac{1-\omega^2 L_R C_L}{\omega C_L}$$

have been used in (24a)–(24c). As expressed in (20a)–(20d), the structural parameters of the CRLH TL unit cell can be used to determine the values of L_R, C_R, L_L and C_L and in turn to determine the phase incursion through (24c) and operation frequency through (24a) and (24b).

4.3 Designs of digital-phase shifters using CRLH unit cells

4.3.1 Basic ideas

The symmetrical CRLH TL unit cell in Fig. 12 (a) is used as the basic cell to design digital-phase shifters here. The advantage of using symmetrical CRLH TL unit cells is that when more unit cells are cascaded together to provide a more states, there is no need to perform any matching between adjacent unit cells.

Here, we use Fig.13 to illustrate our proposed idea of using CRLH TL unit cells to design digital-phase shifters. In the figure, we mount four switches on four different fingers of a unit cell. For an n-bit phase shifter, there will be a total of 2^n states, from 0 to 2^n-1, determined by "closed" or "open" states of the switches on the unit cell. The n-controlling bits are used to select a particular state from these 2^n-1 states and hence to provide a particular phase shift. The "closed" or "open" states in the switches determine the values L_R, C_R, L_L and C_L through (20a)-(20d) and, in turn, determine the phase incursion $Ang(S_{21})$ in (24c), so each switch state can be used to provide a particular phase shift, i.e.,

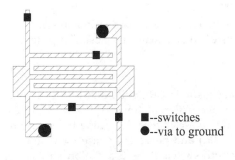

■--switches
●--via to ground

Fig. 13. Symmetrical CRLH TL unit cell mounted with four switches

$$\Delta\theta_m = Ang(S_{21})_m - Ang(S_{21})_0 \tag{25}$$

where m is the index (from 0 to 2^n-1) for the switch states, and $(S_{21})_m$ and $(S_{21})_0$ are the values of S_{21} in the mth-switch and zeroth-switch states, respectively. For convenience and without lost of generality, the zeroth-switch state is taken as the state with all switches closed.

For high power applications such as radars, power-handling capability is one of the important concerns in the design of phase shifters. In our design, the power-handling capacity of the switches (i.e., PIN diodes in our case) used in the CRLH TL unit cell determines the power-handling capacity of the phase shifter. Here, surface-current density distribution is used to study the power-handling capacity of the switches. Computer simulation results on the surface-current density distribution of the CRLH TL unit cell on a Rogers substrate, RO5880, with eight fingers and four switches, are shown in Fig. 14. Figs. 14(a) and 14(b) show the surface-current density distributions with all four switches "opened" or "closed", respectively, at 9.5 GHz. The arrowheads indicate the positions of the switches on the fingers of the CRLH TL unit cell. With all switches "opened", Fig. 14 (a) shows that the largest surface-current density flowing through the switches is about -28 dB below (or 1/25.1 of) those flowing through the input and output ports. While with all switches "closed", Fig. 14 (b) shows that the largest surface-current density flowing through the switches is about -16 dB below (or 1/6.3 of) those flowing through the input and output ports. The width of the finger is only 1/7 of the port, so the largest surface-current through the switches is 16.43 dB below (or 1/44 of) those flowing through the ports. Thus the power-handling capability of the phase shifter is about 44 times higher than the power-handling capability of a switch (PIN diode).

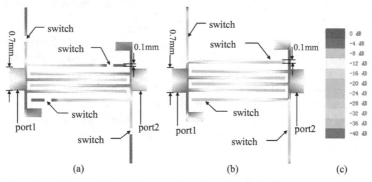

Fig. 14. Simulated surface-current density distribution on symmetrical CRLH TL unit cell with (a) all switches "opened", (b) all switches "closed", and (c) color scale

4.3.2 2-bit phase shifter

A 2-bit phase shifter using a single CRLH TL unit cell is shown in Fig. 15(a), where two switches are mounted on two different fingers (Zhang et al., 2010). The positions of the switches in the "open" and "close" states determine the values of L_R, C_R, L_L and C_L of the unit cell through (20a) – (20d) and hence the phase incursion. The state with all switches closed is taken as the zeroth-switch state. Simulation studies are used to determine the positions of the switches on the top and bottom fingers of the unit cell in order to achieve the phase shifts of 45^0 and 22.5^0, respectively. The phase shifts of 67.5^0 and 0^0 are provided by "opening" and "closing", respectively, both two switches. Table 2 shows the phase shift for different input logic patterns. It should be noted that, although the phase shifts of 0^0, 22.5^0, 45^0 and 67.5^0 with equal step size of 22.5^0 are used in our design, other phase shift values and step sizes are just possible and easily be achieved. The layout of the 2-bit phase shifter with the DC bias circuits for the PINs (the switches) is shown in Fig. 15(b), where RF_{in} and RF_{out} are the input and output RF signals, respectively, and the input bits have the DC voltages of 0 and 5 V representing the two corresponding logic levels "1" and "0".

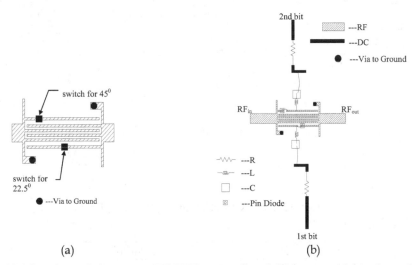

(a)   (b)

Fig. 15. (a) 2-bit phase shifter using CRLH TL unit cell and (b) layout of 2-bit phase shifter with DC bias circuits

States	Phase shifts
00	0^0
01	22.5^0
10	45^0
11	90^0

Table 2. Phase shifts for different states of 2-bit phase shifter

4.3.3 3-bit phase shifter

Fig. 16(a) shows a 3-bit phase shifter using a single CRLH TL unit cell where four switches are used to achieve eight different switch states and hence eight phase shifts of 0^0, 22.5^0, 45^0, 67.5^0, 90^0, 112.5^0, 135^0 and 157.5^0 (Zhang et al., 2010). Note that the number of controlling bits is less than the number of switches on the unit cell because two switches may be used to provide a phase shift. The state with all switches closed is taken as the zeroth-switch state. Again, computer simulation is used to determine the positions of these switches on the fingers in order to achieve these phase shifts. In Fig. 16(a), the two switches on the inner fingers are used to provide the phase shifts of 22.5^0 and 45^0. The two switches on the outer fingers serving in pair are used to provide a total phase shift of 90^0. Different switch states are used to provide other phase shifts such as 67.5^0, 112.5^0, 135^0 and 157.5^0. When all switches are closed, the phase shift is 0^0. Table 3 shows the phase shift for different input logic pattern. The layout of the 3-bit phase shifter with DC bias circuit is shown in Fig. 16(b).

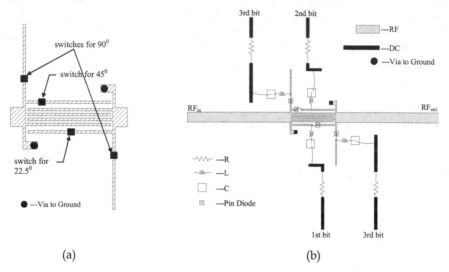

(a)   (b)

Fig. 16. (a) 3-bit phase shifter using one CRLH TL unit cell and (b) layout of 3-bit phase shifter with DC bias circuits

States	Phase shifts
000	0^0
001	22.5^0
010	45^0
011	67.5^0
100	90^0
101	112.5^0
110	135^0
111	157.5^0

Table 3. Phase shifts for different states of 3-bit phase shifter

4.3.4 6-bit phase shifter

To design a 6-bit phase shifter, we employ two symmetrical CRLH TL unit cells, as shown in Figs. 17(a) and 17(b), in cascade to provide the required phase shifts (Zhang et al., 2010). The state with all switches closed is taken as the zeroth-switch state. In the unit cell of Fig. 17(a), we use four switches to achieve sixteen different switch states and hence sixteen different phase shifts. The two switches on the inner fingers are used to provide the phase shifts of 5.625^0 and 11.25^0. The two switches on the outer fingers are used to provide the phase shifts of 22.5^0 and 45^0. Different switch states are used to provide other phase shifts such as 16.875^0, 28.125^0, 34.75^0, 39.375^0, 50.625^0, 56.25^0, 61.875^0, 67.5^0, 74.125^0,

78.75^0 and 84.375^0. Thus these switch states can be used to provide the phase shifts from 0^0 up to 84.375^0 at a step size of 5.625^0. In the unit cell of Fig. 17(b), we use two switches in a pair to provide the large phase shifts of 90^0 or 180^0. The pair of switches on the inner fingers provides a phase shift of 90^0, while the pair of switches on the outer fingers provides a phase shift of 180^0. The phase shift of 270^0 is achieved by "opening" all four switches. This phase shifter therefore can provide the phase shifts from 0^0 up to 270^0 at a step size of 90^0. By cascading these two unit cells, any phase shifts at a multiple number of 5.625^0 can be achieved by using the 6 controlling bits according to (25), i.e. the phase shifter can provide the phase shifts from 0^0 up to 354.375^0 at a step size of 5.625^0. Table 4 shows the phase shift for different input logic pattern. The layout of the 6-bit phase-shifter with DC bias circuits is shown in Fig. 17(c).

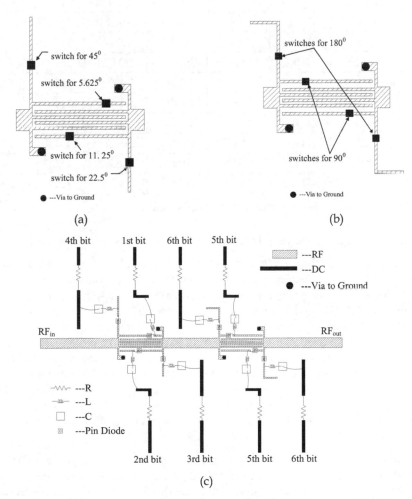

Fig. 17. 6-bit phase shifter using two CRLH TL unit cells. (a) CRLH unit cell in 1st stage, (b) CRLH unit cell in 2nd stage, and (c) layout of 6-bit phase shifter with DC bias circuits

States	Phase shifts	States	Phase shifts	States	Phase shifts	States	Phase shifts
000000	0^0	010000	90^0	100000	180^0	110000	270^0
000001	5.625^0	010001	95.625^0	100001	185.625^0	110001	275.625^0
000010	11.25^0	010010	101.25^0	100010	191.25^0	110010	281.25^0
000011	16.875^0	010011	106.875^0	100011	196.875^0	110011	286.875^0
000100	22.5^0	010100	112.5^0	100100	202.5^0	110100	292.5^0
000101	28.125^0	010101	118.125^0	100101	208.125^0	110101	298.125^0
000110	33.75^0	010110	123.75^0	100110	213.75^0	110110	303.75^0
000111	39.375^0	010111	129.375^0	100111	219.375^0	110111	309.375^0
001000	45^0	011000	135^0	101000	225^0	111000	315^0
001001	50.625^0	011001	140.625^0	101001	230.625^0	111001	320.625^0
001010	56.25^0	011010	146.25^0	101010	236.25^0	111010	326.25^0
001011	61.875^0	011011	151.875^0	101011	241.875^0	111011	331.875^0
001100	67.5^0	011100	157.5^0	101100	247.5^0	111100	337.5^0
001101	73.125^0	011101	163.125^0	101101	253.125^0	111101	343.125^0
001110	78.75^0	011110	168.75^0	101110	258.75^0	111110	348.75^0
001111	84.375^0	011111	174.375^0	101111	264.375^0	111111	354.375^0

Table 4. Phase shifts for different states of 6-bit phase shifter

4.4 Simulation and measurement results

The 2-bit, 3-bit and 6-bit phase shifters in Figs. 15(b), 16(b) and 17(c), respectively, have been designed to operate in an operating frequency band of 9-10 GHz using computer simulation. The substrates used in our designs were Rogers, RO5880, with a thickness of 0.254 mm and a permittivity of 2.2. The switches used were PIN diodes from SKYWORKS Co. Ltd, having a dimension of 0.35 mm×0.35 mm×0.15 mm, operating frequency range of 100 MHz-18 GHz and instantaneous power-handling capability of 2.5 W (34 dBm). Surface-mount-technology components such as resistors, capacitors and inductors were used to construct the bias circuits for the PIN diodes and also the circuits to isolate the RF signal from DC. For verification, all the final designs have been fabricated using Rogers substrate, RO5880, the same substrate used in our simulations. The prototyped modules of the 2-bit, 3-bit and 6-bit phase-shifters are shown in Fig. 18, having the dimensions of 35 mm×40 mm×2 mm, 50 mm×40 mm×2 mm and 60 mm×40 mm×2 mm, respectively.

The return losses ($-10\log|S_{11}|$), insertion losses ($-10\log|S_{21}|$) and phase shifts of these prototyped phase shifters have been measured using a network analyzer to verify the simulation results. The simulation and measurement results for the 2-bit, 3-bit and 6-bit phase shifters are shown in Figs. 19, 20 and 21, respectively. It can be seen that the simulated and measured results show good agreements.

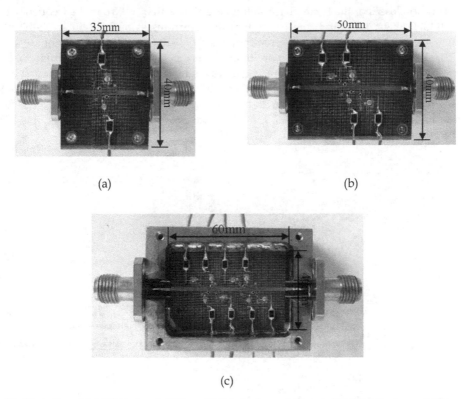

(a)

(b)

(c)

Fig. 18. Prototyped (a) 2-bit phase shifter, (b) 3-bit phase shifter and (c) 6-bit phase shifter

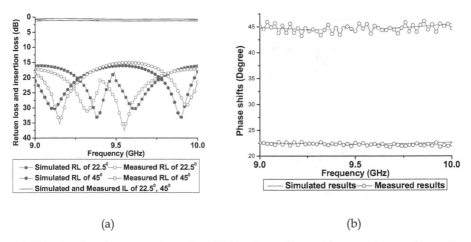

(a)

(b)

Fig. 19. Simulated and measured results of 2-bit phase shifter; (a) return loss and insertion loss and (b) phase shift

The insertion losses of the three phase shifters are all less than 1.3 dB in the frequency band of 9-10 GHz. Figs. 19(a) and 20(a) show that the return losses of the 2-bit and 3-bit phase shifters are more than 15 dB in the frequency band of 9-10 GHz. For the 6-bit phase shifter, Fig. 21(a) shows the return loss is more than 15 dB for the frequency band from 9.2 to 9.8 GHz. Regarding accuracy, the measurement results in Fig. 19(b) show that the maximum error for the 2-bit phase shifter is about -2^0 at the phase shift of 67.5^0 and frequency of 9.73 GHz. For the 3-bit phase shifter, Fig. 20(b) shows that the maximum errors is $+4.5^0$ at the phase shift of 157.5^0 and frequency of 9.10 GHz. While for the 6-bit phase shifter, it is -6.3^0 at the phase shift of 180^0 and frequency of 9.66 GHz.

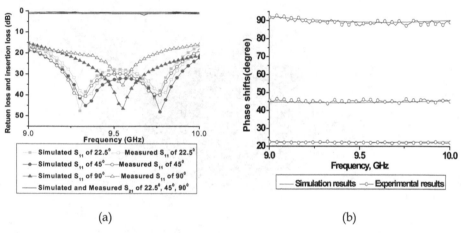

(a)   (b)

Fig. 20. Simulated and measured results of 3-bit phase shifter; (a) return loss and insertion loss and (b) phase shift

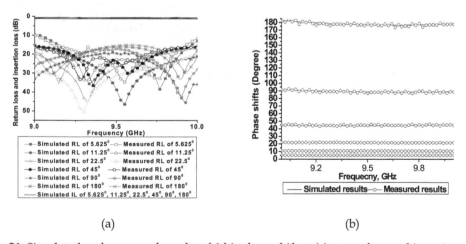

(a)   (b)

Fig. 21. Simulated and measured results of 6-bit phase shifter; (a) return loss and insertion loss and (b) phase shift

The power-handling capability of the phase shifters has been studied using a setup with the block diagram shown in Fig. 22. The solid-state transmitter is basically an AM modulator which takes in the signal from the IF signal source, uses it to modulate the carrier signal at a much higher frequency and amplifies the modulated signal to a high-power level. The amplified signal is fed to the phase shifter under tested via an isolator which prevents any high-power signal from reflecting back and damaging the solid-state transmitter. A 40-dB coupler is used to couple a small portion of the high-power signal from the output of the phase shifter and feed it to the frequency spectrograph for measuring the power capability of the phase shifter. Majority of the high-power signal from the phase shifter is fed to a high-power load for power dissipation.

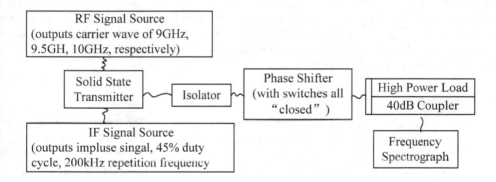

Fig. 22. Equipment setup for measuring power-handling capability of phase shifter

Frequency (GHz)		9	9.5	10
	2-bit	51.2	51.3	51.1
Peak Power (dBm)	3-bit	51.1	51.2	51.4
	6-bit	51.3	50.9	51.1

Table 5. Power-Handling Capability of Phase Shifters

In this study, the maximum transmitted peak power causing a diode to burn off is taken as the power-handling capability of the phase shifter and the results are shown in Table 5. Since the PIN diodes used in our design have an instantaneous power-handling capability of 34 dBm, the expected power-handling capabilities for the 2-bit, 3-bit and 6-bit phase shifters are about 50.43 dBm, i.e. 44 times higher as described previously. Table 4.4 shows that the power-handling capabilities of the three phase shifters at 9, 9.5 and 10 GHz are over 51.1, 51.1 and 50.9 dBm, respectively, corresponding to about 51.5, 51.5 and 49.2 times of the power-handling capability of the PIN diodes.

5. Conclusions

In this chapter, the EM metamatreials and the realization of EM metamaterials using the transmission line approach have been briefly described. A CRLH TL unit cell with a symmetrical structure has been proposed for the designs of TTDLs and digital-phase shifters. TTDLs using a single unit cell and four unit cells in cascaded have been designed and studied. Simulation and measurement results have shown that the TTDLs have the return losses of more than 15 dB and insertion losses of less than 1 dB. For the same length of 30 mm, the TTDL constructed using four CRLH TL unit cells in cascade can achieve a much larger TD, about 3.2 times larger, than that using the RH TL. Digital-phase shifters constructed using the CRLH TL unit cells have the advantages of small sizes, arbitrary phase-shift ranges and arbitrary step sizes. Three digital-phase shifters, 2-bit, 3-bit and 6-bit, have been designed and studied. Simulation and experimental results have shown that the digital-phase shifters have the low insertion losses of about 1.3 dB and return losses of larger than 15 dB across the operation bandwidths. Moreover, they have the much higher power-handling capabilities, about 50 times higher, than that of the PIN diodes used as switches in our designs for the digital-phase shifters.

6. References

Adam, J.D.; Davis, L.E.; Dionne, G.F.; Schloemann E.F. & Stitzer, S.N. (2002). Ferrite devices and materials. *IEEE Transactions on Microwave Theory and Techniques*, Vol. 50, 2002, pp. 721-737

Antoniades, M.A. & Eleftheriades, G.V. (2003). Compact linear lead/lag metamaterial phase shifters for broadband applications. *IEEE Antennas and Wireless Propagation Letters*, Vol. 2, 2003, pp.103-106

Bahl, I. (2011). *Lumped Elements for RF and Microwave Circuits*, Artech House, Boston& London

Caloz, C. & Itoh, T. (2006). *Electromagnetic Metamaterials: Transmission Line Theory and Microwave Applications: the Engineering Approach*, John Wiley & Sons, New Jersey

Damm, C.; Schüßler, M.; Freese J. & Jakoby, R. (2006). Artificial line phase shifter with separately tunable phase and line impedance, *36th European Microwave Conference*, 2006

Fetisov, Y.K. & Kabos, P. (1998). Active Magnetostatic Wave Delay Line. *IEEE Transactions on Magnetics*, Vol. 34, 1998, pp. 259 – 271

Keul, S. & Bhat, B. (1991). *Microwave and millimeter wave phase shifters*, Artech house, MA

Kim, H.; Kozyiev, A.B.; Karbassi, A. & van der Weide, D.W. (2005). Linear Tunable Phase Shifter Using a Left-handed Transmission-line. *IEEE Microwave and Wireless Components Letters*, Vol. 15, 2005, pp. 366-368

Kholodnyak, D.V.; Serebryakova, E.V.; Vendik, I.B. & Vendik, O.G. (2006). Broadband Digital Phase Shifter Based On Switchable Right- and Left-handed Transmission Line Sections. *IEEE Microwave and Wireless Components Letters*, Vol. 1, 2006, pp. 258-260

Kim, H. & Drayton, R.F. (2007). Size Reduction Method of Coplanar Waveguide (CPW) Electromagnetic Bandgap (EBG) Structures Using Slow Wave Design, *Topical Meeting on Silicon Monolithic Integrated Circuits in RF Systems*, 2007

Lai, A.; Itoh, T. & Caloz, C. (2004). Composite right/left-handed transmission line metamaterials. *IEEE Microwave Magazine*, Vol. 5, Sep 2004, pp. 34 – 50

Lee, T.H. (2004). *Planar Microwave Engineering*, Cambridge University Press, UK

Lapine, M.; Nefedov, I.S.; Saeily, J. & Tretyakov, S.A. (2006). Artificial lines with exotic dispersion for phase shifters and delay lines, *36th European Microwave Conference*, 2006

Marc, E.G. & Robert, A.P. (1991). Modelling Via Hole Grounds in Microstrip. *IEEE Microwave and Guided Wave Letters*, Vol. 1, 1991, pp. 135-137

Monti, G.; DePaolis, R. & Tarricone, L. (2009). Design of a 3-state reconfigurable CRLH transmission line based on MEMS switches, *Progress In Electromagnetics Research*, PIER 95, 2009, pp. 283-297

Pozar, M.D. (2004). *Microwave Engineering*, John Wiley & Sons, New Jersey

Rebeiz, G.M.; Tan, G.L. & Hayden, J.S. (2002). RF MEMS phase shifters: design and applications. *IEEE Microwave Magazine*, Vol. 2, 2002, pp. 72-81

Smith, W.R. & Gerard, H.M. (1969). Design of Surface Wave Delay Lines with Interdigital Transducers. *IEEE Transactions on Microwave Theory and Techniques*, Vol. 17, 1969, pp. 865 – 873

Ulaby, F.T. (2004). *Fundamentals of Applied Electromagnetics*, Pearson/Prentice Hall, New Jersey

Veselago, V.G. (1968). The electrodynamics of substances with simultaneously negative values of ε and μ, *Soviet Physics Uspekhi*, January-Feburary 1968

Vendik, I.B.; Kholodnyak, D.V. & Kapitanova, P.V. (2009). Microwave phase shifters and filters based on a combination of left-handed and right-handed transmission lines, In: *Metamaterials Handbook Vol. II. Applications of Metamaterials*, pp. 1-21, CRC Press

Woo, D.J.; Lee, J.W. & Lee, T.K. (2008). Multi-band Rejection DGS with Improved Slow-wave Effect, *38th European Microwave Conference*, 2008

Xu, J.; Lu, Z. & Zhang, X.C. (2004). Compact Involute Optical Delay Line. *Electronic Letters*, Vol. 40, 2004, pp. 1218 – 1219

Zhang, Y.Y. & Yang, H.Y.D. (2008). Ultra slow-wave periodic transmission line using 3D substrate metallization, *Microwave Symposium Digest 2008 IEEE MTT-S International*, 2008

Zhang, J.; Zhu, Q.; Jiang, Q. & Xu, S.J. (2009). Design of time delay lines with periodic microstrip line and composite right/left-handed transmission line. *Microwave and Optical Technology Letters*, Vol. 51, 2009, pp. 1679-1682

Zhang, J.; Cheung, S.W. & Yuk, T.I. (2010). Design of n-bit Phase Shifters with High Power Handling Capability Inspired by Composite Right/Left-handed Transmission Line Unit Cells. *IET Microwaves, Antennas & Propagation*, Vol. 4, August 2010, pp. 991-999

Zhang, J.; Cheung, S.W. & Yuk, T.I. (2010). A 3-bit Phase Shifter with High-power Capacity Based on Composite Right/Left-handed Transmission Line. *Microwave and Optical Technology Letters*, Vol. 52, August 2010, pp.1778-1782

Zhang, J.; Cheung, S.W. & Yuk, T.I. (2011). UWB True-time-delay Lines Inspired by CRLH TL Unit Cells. *Microwave and Optical Technology Letters*, Vol. 53, September 2011, pp.1955-1961

4

Perfect Metamaterial Absorbers in Microwave and Terahertz Bands

Qi-Ye Wen*, Huai-Wu Zhang, Qing-Hui Yang, Zhi Chen, Bi-Hui Zhao, Yang Long and Yu-Lan Jing

State Key Laboratory of Electronic Films and Integrated Devices, University of Electronic Science and Technology of China, Chengdu, China

1. Introduction

Recently, resonant metamaterial absorbers (MAs) at microwave and terahertz (THz) bands have attracted much attention due to the advantages such as high absorption, low density, and thin thickness [1–6]. The MA generally composed of a metamaterial layer and a metal plate layer separated by a dielectric spacer. With this kind of novel device, unity absorptivity can be realized by matching the impedance of MA to free space. Besides that, wide-angle, polarization insensitive and even multi-bands/wide-band absorption can be achieved through properly device designing [7-11]. Furthermore, our previous investigations on MA show that the absorber traps the incident electronmagnetic (EM) wave into some specific spots of the devices, and then converts it into heat [9, 12]. All these features make MAs very useful in areas such as EM detector/imager, anti-electromagnetic interference, stealth technology, phase imaging, spectroscopy and thermal emission.

Tunable devices, which allow one to real-time control and manipulate of EM radiation, are emerging as an interesting issue in metamaterials fields [13]. Combined with pin diodes, switchable microwave MAs were developed and an electronic control strategy was demonstrated [14, 15]. However, the device fabrication process is complex, and the structure with pin diodes is hard to be scaled down to higher frequency such as THz and visible regimes. In this chapter, we proposed a VO_2 based switchable MA in microwave band. VO_2 is known to exhibit a transition from an insulating phase to a metallic state (IMT) when it is thermally, electrically or optically triggered. By this unique property, VO_2 films have already been used to tune the resonance characteristics of metamaterial in near IR [16] and THz regimes [17, 18]. The switchable microwave MA presented in this work is realized by placing VO_2 thin film between the electronic split ring resonator (eSRR) and the dielectric layer. It is found that by triggering the IMT of the VO_2, the absorption amplitude of the device can be significantly switched between absorber and reflector with high speed. This VO_2 based MA has the advantages such as simple fabrication, strong tunability and easy to scale to terahertz and optical bands. This tunable MA also has the potential to be a self-

---

*Corresponding Author

resetting "smart" EM absorber, since the absorbed wave would transfer into heat and that can also trigger the IMT of the VO_2 film.

2. Basic structure and working principle of metamaterial absorber

The first metamaterial based absorber was proposed by N. L. Landy et al in the microwave band [1]. It is called "perfect metamaterials absorber" because nearly 100% absorption can be achieved theoretically. A single unit cell of the absorber consisted of three layers as shown in Fig.1. The top layer is the electric split-ring resonator (eSRR), the middle layer is isolation layer (such as polyimide), and the bottom layer is rectangular metal strip. Due to the lithography alignment and multi-step lithography process, the preparation process of the first MA is complex. Experimentally, the maximum absorptivity only reaches to 70% at 1.3THz due to the fabrication tolerance. An improved MA was proposed by H. Tao et al, with the bottom metal strips replaced by a continuous metal film [2]. This improved absorber operates quite well for both TE and TM radiation over a large range of incident angles (0-50°), and the measured absorbance was further improved to 97%. Therefore, the MA with a continuous metal ground plane become the most commonly structure in the researches.

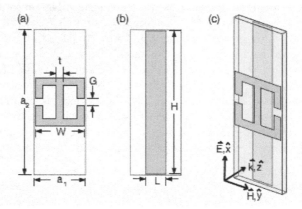

Fig. 1. Basic structure of three layers metamaterial absorber (a)the layer of metamaterial;(b) the bottom metallic layer;(c)the unit structure include the dielectric layer. Figure adapted from N. L. Landy et al. [1]

The absorption mechanism of the MA is as follows: First, by changing the geometry of the SRR and the thickness of the spacer, the impedance $Z(\omega)$ of the absorber can be designed to match the impedance of free space at a specific frequency (center frequency) resulting in zero reflection; Second, electromagnetic waves can not pass through the metallic ground plane, giving rise to zero transmission too. Thus, electromagnetic waves will be completely restricted in the device and finally be consumed. In principal, the metamaterial absorber can absorb 100% of the narrow-band electromagnetic waves. It can be used in microwave, terahertz (THz) and even light wave band by adjusting the feature size of the unit cell.

Though more and more attentions have been paid to MA, the mechanism of the near-unity absorption is still under studying. It has been suggested the matching between the effective permittivity and permeability may be able to interpret the perfect absorption. However, the effective medium theory has some problems in describing MA because the three-layer structured device doesn't exactly satisfy the homogeneous-effective limit, according to Caloz [20]. A typical case is that the strong asymmetric absorption phenomenon cannot be fully explained by effective medium model [21]. The simulation results show dramatically different behaviors when the electromagnetic waves incident from the two opposite directions. For example, when light is incident from the front to the resonators the device acts as a perfect absorber, while when light is incident from the back to the ground plane the device behaves like a perfect mirror. Furthermore, the MA consists of only two metallic layers, thus are strongly inhomogeneous in the wave propagating direction, which is obviously in contrast to the effective medium model.

Q.Y. Wen et al have proposed a transmission line (TL) mode based on the equivalent RLC model [12]. In the TL model, it is assumed that the transverse electromagnetic (TEM) wave propagates through free space and the substrate with intrinsic impedances Z_i and Z_o respectively. There are two assumptions for constructing the TL model. One is that coupling capacitor or coupling inductor between the eSRR layer and wires layer should be ignorable, so that these two layers can be individually modeled, as demonstrated in Fig. 1. Another is that the THz wave normally incidents on the absorber plane with the electrical field parallel to the split gap of the eSRR. The TL model of eSRR proposed by A. K. Azad [22] is used to describe the eSRR layer, in which the LC resonance and dipole resonance each is represented by one group of L, C and R respectively, and the coupling between these two resonances is specified by the parameter M. The wires layer part is mimicked by the TL model developed by L. Fu [23], with the only resonance expressed by one group of L, C and R. The function of isolation layer is modeled by a transmission line which contains all EM related properties of the isolation layer such as ε, μ and thickness. It connects the eSRR part and wire part. All the parameters are needed to be optimized until the S-parameters calculated by the TL model fit the simulation results.

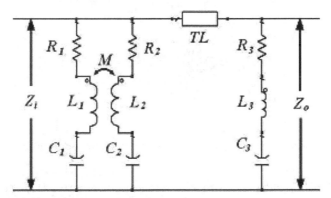

Fig. 2. Transmission line model of metamaterial absorber

By this TL model, the asymmetric phenomenon of THz absorption is unambiguously demonstrated and explained. The strong absorption is found to be mainly related to the LC resonance of the eSRR structure. The isolation layer in the absorber, however, is actually an impedance transformer and plays key role in producing the perfect absorption. The studies by TL model also show that the electromagnetic wave is concentrated on some specific location in the absorber. It indicates that the trapped electromagnetic wave in the absorber can be converted into thermal energy, electric energy or any kinds of other energy depending on the functions of the spacer materials. This feature as electromagnetic wave trapper has many potential applications such as radiation detecting bolometers and thermal emitter.

3. Progress in metamaterial absorber

The perfect absorber first proposed by N. L. Landy is an anisotropic absorber. They further proposed a polarization independent absorber, as shown in Fig.3 [5]. One of the main features of the structure is that the eSRR has fourfold-rotational symmetry about the propagation axis and was therefore polarization insensitive. The measured absorptivity of this absorber is 65% at 1.145 THz. Experimental results confirmed that the polarization-independent metamaterial absorber can be realized by chosen a fourfold rotational symmetry SRR structure.

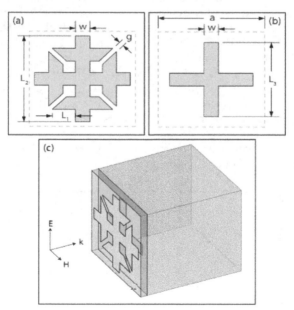

Fig. 3. Polarization-independent metamaterial absorber. Figure adapted from N. I. Landy et al. [5]

H. Tao et.al designed a wide-angle terahertz metamaterial absorber [13]. Most importantly, the device was fabricated on a highly flexible polyimide substrate with a total thickness of 16 um. This novel design enables its use in nonplanar applications as it can be easily wrapped around objects as small as 6 mm in diameter. They demonstrated, through simulation and experiment, that this metamaterial absorber operates over a very wide range of angles of incidence for both transverse electric (TE) and transverse magnetic (TM) configuration.

Apart from single-frequency absorber, dual-band and multi-band absorbers also draw attention from researchers. A dual-band metamaterial absorber was demonstrated in terahertz band, as shown in Fig 4 [9]. The special feature of this absorber is that its eSRR unit possesses two kinds of split gaps and therefore exhibits two well-separated LC resonances. Theoretical calculation shows that there are two distinct absorptive peaks located around 0.50 and 0.94THz, each with absorptions over 99.99%. The measured absorption is 81% for low frequency absorption and 63.4% for high frequency absorption. The experimental result suggests that the design of an eSRR with multiple LC resonances is a key step toward a multi-frequency absorber. A similar result was also reported by H. Tao et al [10].

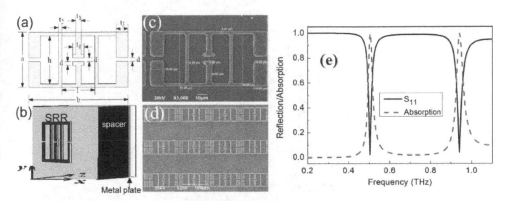

Fig. 4. Dual-band THz metamaterial absorber. (a) Designed electric split ring resonator, (b) Perspective view of the designed absorber. (c) A unit cell of the experimentally realized absorber. (d) Photograph of a portion of the fabricated absorber. (e) The simulated reflection (solid line) and absorption (dotted line) curve of the absorber.

X. P. Shen et al. have developed a wide angle triple-band absorber structure very recently, as show in Fig.5 [11]. The top layer consists of an array of three nested copper closed ring resonator arrays, which is primarily responsible for the electric response to the incident field. The bottom layer is a copper plane, which is used to zero the transmission and is responsible for the magnetic response. The experimental results show three absorption peaks at frequencies 4.06GHz, 6.73GHz and 9.22GHz with absorptivity of 99%, 93%, and 95%, respectively, which agrees well with the simulation results.

In certain applications, broadening the absorption bandwidth is also of importance. Fig.6 shows an omnidirectional polarization-insensitive absorber with a broadband feature in the terahertz regime is proposed by Q.Y. Ye et al. [8]. They demonstrated that the bandwidth of the absorption can be effectively improved by using a multilayer structure, while the wide-angle feature remains. A simple cross-shaped pattern was used as the resonator. The experiment results show that with the increasing of metamaterial layers, the absorption peak is gradually broadened. A bandwidth of nearly 1000 GHz with perfect absorption (more than 97%) was achieved in the three layer absorber.

A frequency tunable metamaterial absorber is proposed by incorporated a pin-diode between two resonators [15]. Simulation and measurement results show that by forward or reverse biasing the diodes so as to change the coupling between the resonators, the absorber can be dynamically switched to operate in two adjacent frequency bands with nearly perfect peak absorption. It is also shown that by tuning the loading position of the diodes, it is able to adjust the frequency difference between the two switchable absorbing bands.

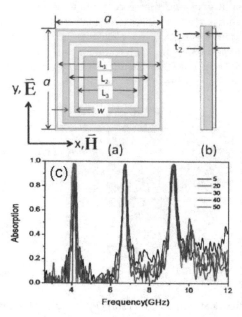

Fig. 5. (a) The front and (b) side view of the unit cell of the triple band microwave absorber, and (c) simulation and experimental results of the triple-band absorber at various angles of incidence. Figure adapted from X. P. Shen ea al [11].

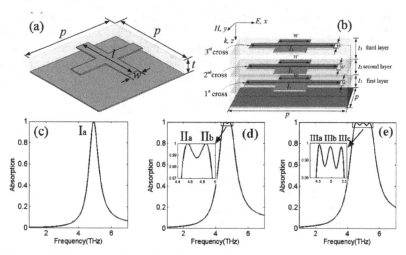

Fig. 6. (a) Unit cell of the absorbers, (b) Schematic diagram of a 3-layer cross structure, and bsorption spectra for (c) 1-layer cross structure; (d) 2-layer cross structure; (e) 3-layer cross structure.

4. Switchable metamaterials absorber based on vanadium dioxides

Though tunable metamaterial absorber has been demonstrated in microwave band, the device fabrication process is complex, and the structure with pin diodes is hard to be scaled down to higher frequency such as THz and visible regimes. In this chapter, we proposed a novel switchable MA in microwave band, which is simple and can be tuned thermally, electrically and optically. The MA under studied has a similar structure to our previously designed dual band THz absorber [9] except that a thin VO_2 film is incorporated between the eSRR and the substrate, as shown in Fig.7. The eSRR unit cell is composed of two kinds of SRR with one (the inner resonator) invaginated in the other (the outer resonator pairs). It should be noted that, for the purpose of comparison, only the inner resonator has an underlying VO_2 patch and the outer resonators contact to the substrate directly. As shown in Fig.7 (b), the area of the VO_2 patch is equal to the inner resonator. Utilizing the commercial software CST Microwave Studio ™ 2009, the size parameters of the eSRR marked in Fig.7 (a) was optimized to obtain two strong absorptions. In the simulation, the microwave MA is built on C-cut sapphire substrate with one side covered with metallic ground plane and another side with eSRR arrays. Both of the metallic eSRR and ground plane are modeled as copper sheet with conductivity of 5.8×10^7 S/m. A 200nm thick VO_2 film was applied in the simulation. The permittivity and the permeability of VO_2 were set to be 3 and 1, respectively. The conductivity (σ) was swept from 0.02S/m to 2000S/m to mimic the phase transition process. The incident microwave wave is normal to the devices plane with the electronic field perpendicular to the split gaps, e.g. along y axis as indicated in Fig.7 (b). The optimized parameters are: a=2mm, b=4mm, d=0.1mm, h=1.8mm, l=1mm, t1=t2=0.2mm, t3=0.3mm, t4=0.5mm and the unit cell is 3mm × 5mm. The copper thickness of eSRR layer is 0.2μm and that for ground plane is 0.8μm.

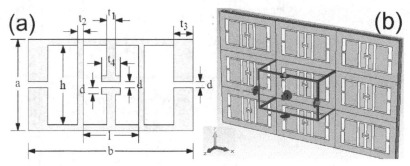

Fig. 7. Dual-band microwave metamaterial absorber. (a) Schematic of the designed eSRR unit cell. (b) Perspective view of the absorber sheet with VO_2 patterns. Gray represents the sapphire substrate, yellow the copper and dark green the VO_2 film.

Fig.8 shows the simulated S_{11} parameters of the MA with different conductivity of the VO_2 patch. When σ is 0.02S/m, e.g, the insulating state of the VO_2 film, two distinct reflection peaks are observed around 9.03 GHz and 17.6 GHz. Since the transmission of MA is zero due to the metallic bottom plane, the absorptivity can be calculated using $A = 1 - |S_{11}|^2$. High absorptivity of 92.7% and 99.4% was obtained at 9.03 GHz and 17.6 GHz, respectively. The surface current distributions on the top eSRR and bottom plane layer at resonance frequencies were monitored, as shown in Fig.9. The results confirmed that both resonances come from circulating currents in the eSRR resonators [9, 10]. Furthermore, it shows that the low frequency response is determined mainly by the outer ring pairs (Fig 9 (a) & (c)), while the high frequency response is induced mainly by inner resonator (Fig.9 (b) & (d)).

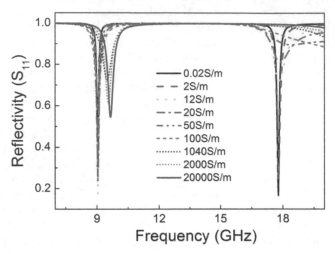

Fig. 8. Simulated reflection curve for metamaterial absorber with different conductivity of VO₂

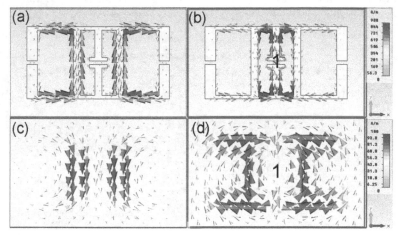

Fig. 9. Surface current distributions in the top and bottom metallic layers. (a) and (c) for the low frequency absorption, and (b) and (d) for high frequency absorption

When the value of σ increases to 2S/m, the high frequency reflectivity begins to increase. With the increase of σ from 2S/m to 50S/m, the peak reflectivity around 17.6GHz increases abruptly from 0.25 to 0.90, corresponding to a decrease of absorptivity from 93.75% to 19%. Further increase of σ gives rise to a slight increase of the reflectivity. As to the low frequency absorption, however, the situation is different. It seems that the variation of σ from 0.02S/m to 20S/m have little effect on absorption. Further increase σ from 20 S/m to 20000 S/m leads to a moderate increase of peak reflectivity from 0.26 to 0.54. Interestingly, for both absorptions a blue-shift about 0.6GHz of the peak frequency is clearly observed. All these results indicate that the IMT of VO_2 does affect the absorption of the MA, and the mechanism will be discussed later in detail.

With the optimized size parameters the microwave MA device was fabricated. Firstly, a VO_2 layer of almost 200 nm in thickness was deposited on 0.5mm-thick C-type sapphire substrates by using the reactive magnetron sputtering technique. The phase transition temperature (TP) of the films was measured to be around 340K and the conductivity exhibits nearly three-orders decrease [18]. The VO_2 film was then etched using CF_4/O_2 plasma to 1.2mm × 2mm patches in a period of 3mm × 5mm, which is the identical period of the eSRR array. After that copper film with thickness of 0.2μm was sputtering deposited on top side of the VO_2/Substrate, following by a conventional lithography process to form the designed eSRR structure with its inner resonator overlapping on the VO_2 patch. Finally another 0.8μm copper film was deposited on the backside of the substrate.

Based on the principle of Arc Tracking Test method, a vector network analyzer (Agilent 8720ES) with two horn antennas was used to transmit EM waves onto the sample sheet and receive the reflected signals. The incident and receive angle is less than 5° from normal in the experiment. Fig.10 shows the measured S_{11} curves of the VO_2 based MA under different temperature. Two strong reflection peaks are clearly appeared and they are distinct from each other, which agrees with the simulation well. At room temperature (RT), two reflective minimums were appeared at 9.36GHz and 18.6GHz, respectively. By $A=1-|S_{11}|2$ the peak absorptivity at low and high frequency were calculated to be 84.8% and 92.1%. For both absorptions, the measured peak frequencies and absorption amplitudes have slight derivations from the theoretical results, which are probably induced by the fabrication tolerance or the difference of the material parameter between simulation and experiment. Variations of the absorption characteristic of the MA were observed when the device temperature increased from RT to 345K. For the low-frequency absorption, the reflectivity minimum shows a small increase from 0.39 to 0.57, corresponding to an absorption decrease from 84.8% to 67.5%. Significant change of the absorption was happened in the high frequency case. With the increase of temperature from RT to 345K, the reflectivity minimum increases notably from 0.28 to 0.83, accompanying with a blue-shift of the peak frequency form 18.6 GHz to 19.12 GHz. Therefore, a deep amplitude modulation about 60% to the microwave absorption and a moderate frequency shift of about 0.5 GHz was realized. All these results agree with the simulated results very well, confirming that the VO_2 based microwave MA is thermally tunable.

In order to further clarify the tunability of MA, the temperature dependence of the reflectivity amplitude and corresponding peak frequency for high-frequency absorption was summarized and plotted in Fig.10 (b). It can be seen that the violent variation of the absorption characteristics, including the amplitude and peak frequency, occurs during a very narrow temperature range from 337K to 345K, which is the phase-transition temperature range of the VO_2 films under studied. Combining the theoretical and experimental results, we can conclude that it is the thermally triggered IMT of VO_2 that induces the tunability of the MA. To understand how the IMT of VO_2 affects the properties of MA, the distribution of the absorption at low and high frequency was examined at RT and compared in Fig.11. In the low-frequency case, the absorption mainly takes place in the vicinity of the two outer gaps, with a small portion occurring at the four corners and inner split gap. In the high-frequency case, almost all the absorption occurs at inner gap. Therefore, the working principles of our tunable microwave MA can be explained as follows: At RT, the VO_2 is in its insulating state and allows dual band absorptions. When the temperature increase beyond T_P, the VO_2 film changes into the metallic state and electronically shorts the inner split gap of the eSRR. The impedance match for the high frequency absorption is seriously broken thus the absorption is strongly attenuated. For low frequency absorption, though the absorption from the inner resonator is also attenuated as temperature increasing, the absorption from the outer resonator pairs is still strong since there is no underlying VO_2 film. That's why the low frequency absorption is not seriously crippled. The resonance frequency of eSRR can be described as $\omega=(LC)^{-1/2}$, where L and C represents the inductance and capacitance of the resonator [19]. Therefore, the blue-shift of the peak frequencies can be ascribed to the decrease of inductance as a result of the metallization of VO_2 film. These results confirmed that by triggering the IMT of the VO_2 films both the amplitude and frequency of the absorption can be tuned.

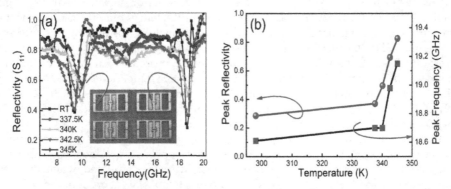

Fig. 10. (a) Measured reflectivity curves of the VO_2 based MA with respect to the devices temperature. Inset is the image of the fabricated device; (b) the temperature dependence of the reflection amplitude and corresponding peak frequency for high frequency absorption.

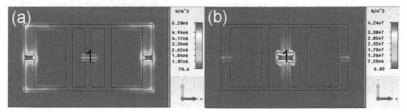

Fig. 11. Distributions of absorption densities for (a) low frequency absorption and (b) high frequency absorption.

5. Summary

Metamaterial absorbers with strong absorbance in microwave and terahertz band are very useful in many regimes such as detecting THz, creating thermal radiation, or cloaking. The metamaterial absorber could be designed to be polarization independent, broadband and even tunable with a small volume, small thermal mass, and high absorption coefficient. In this chapter, a tunable microwave absorber was demonstrated theoretically and experimentally. It is found that by thermally triggering the IMT of the VO_2, the absorption amplitude of the device can be significantly tuned. This tunable MA also has the potential to be a self-resetting "smart" EM absorber, since the absorbed wave would transfer into heat and that can also trigger the IMT of the VO_2 film.

6. Acknowledgement

This work is supported by NSFC under Grant No. 61131005 and 61021061, National Basic Research Program of China (973) under Grant No. 2007CB310407, the "New Century Excellent Talent Foundation" under Grant No. NCET-11-0068, Sichuan Youth S & T foundation under No. 2011JQ0001, and Sichuan International S&T Cooperation Program under Grant No. 2010HH0026. This work is partly supported by "the Fundamental Research Funds for the Central Universities" under Grant No. ZYGX2010J034.

7. References

[1] N. I. Landy, S. Sajuyigbe, J. J. Mock, D. R. Smith, and W. J. Padilla, "Perfect Metamaterial Absorber," Phys. Rev. Lett. 100, 207402 (2008).

[2] H. Tao, C. M. Bingham, A. C. Strikwerda, D. Pilon, D. Shrekenhamer, N. I. Landy, K. Fan, X. Zhang, W. J. Padilla and R. D. Averitt, "Highly flexible wide angle of incidence terahertz metamaterial absorber: Design, fabrication, and characterization," Phys. Rev. B 78, 241103R (2008)

[3] H. Tao, N. I. Landy, C. M. Bingham, X. Zhang, R. D. Averitt, and W. J. Padilla, "A metamaterial absorber for the terahertz regime: Design, fabrication and characterization," Opt. Express 16, 7181-7188 (2008).

[4] Y. Avitzour, Y. A. Urzhumov, and G. Shvets, "Wide-angle infrared absorber based on a negative-index plasmonic metamaterial," Phys. Rev. B 79, 045131 (2009).

[5] N. I. Landy, C. M. Bingham, T. Tyler, N. Jokerst, D. R. Smith and W. J. Padilla, "Design, theory, and measurement of a polarization-insensitive absorber for terahertz imaging," Phys. Rev. B 79, 125104 (2009).

[6] R. Huang, Z. W. Li, L. B. Kong, L. Liu, and S. Matitsine, "Analysis and design of an ultra-thin metamaterial absorber," Progress In Electromagnetic Research B 14, 407-429 (2009).

[7] B. Wang, T. Koschny, and C. M. Soukoulis, "Wide-angle and polarization-independent chiral metamaterial absorber," Phys. Rev. B, 80, 033108 (2009).

[8] Y. Q. Ye, Y. Jin, and S. L. He, "Omnidirectional, polarization-insensitive and broadband thin absorber in the terahertz regime," J. Opt. Soc. Am. B 27, 498-504 (2010)

[9] Q. Y. Wen, H. W. Zhang, Y. S. Xie, Q. H. Yang, and Y. L. Liu, "Dual Band Terahertz Metamaterial Absorber: Design, fabrication, and characterization," Appl. Phys. Lett. 95, 241111 (2009)

[10] H Tao, C M Bingham, D Pilon, Kebin Fan, A C Strikwerda, D Shrekenhamer, W J Padilla, Xin Zhang and R D Averitt, "A dual band terahertz metamaterial absorber," J. Phys. D: Appl. Phys. 43, 225102 (2010)

[11] X. P. Shen, T. J. Cui, J. M. Zhao, H. F. Ma, W. X. Jiang, and H. Li, "Polarization-independent wide-angle triple-band metamaterial absorber," Opt. Express 19, 9401-9407, (2011)

[12] Q. Y. Wen, Y. S. Xie, H. W. Zhang, Q. H. Yang, Y. X. Li, and Y. L. Liu, "Transmission line model and fields analysis of metamaterial absorber in the terahertz band," Opt. Express 17, 20256–20265 (2009)

[13] H. T. Chen, W. J. Padilla, J. M. O. Zide, A. C. Gossard, A. J. Taylor, and R. D. Averitt, "Active terahertz metamaterials devices," Nature, 444, 597 (2006)

[14] A. Tennant. and B. Chambers, "A single-layer tunable microwave absorber using an active FSS," IEEE Microw. Wirel. Compon. Lett., 14, 46-47 (2004)

[15] B. Zhu, C. Huang, Y. Feng, J. Zhao, and T. Jiang, "Dual band switchable metamaterial electromagnetic absorber," Progress In Electromagnetics Research B, 24, 121-129 (2010)

[16] M. J. Dicken, K. Aydin, I. M. Pryce, L. A. Sweatlock, E. M. Boyd, S. Walavalkar, J. Ma, and H. A. Atwater, "Frequency tunable near-infrared metamaterials based on VO_2 phase transition," Opt. Express 17, 18330-18339 (2009)

[17] T. Driscoll, S. Palit, M. M. Qazilbash, M. Brehm, F. Keilmann, B. G. Chae, S. J. Yun, H. T. Kim, S. Y. Cho, N. M. Jokerst, D. R. Smith, and D. N. Basov, "Dynamic tuning of an infrared hybrid-metamaterial resonance using vanadium dioxide," Appl. Phys. Lett. 93, 024101 (2009)

[18] Q. Y. Wen, H. W. Zhang, Q. H. Yang, Y. S. Xie, K Chen, and Y. L. Liu, "Terahertz Metamaterials with VO_2 Cut-wires for Thermal Tunability," Appl. Phys. Lett. 97, 021111 (2010)

[19] J. Q. Gu, J. G. Han, X. C. Lu, R. Singh, Z. Tian, Q. R. Xing, and W. L. Zhang, "A close-ring pair terahertz metamaterial resonating at normal incidence," Opt. Express 17, 20307-20312 (2009).

[20] A. C. Caloz, and T. Itoh. "Electromagnetic Metamaterial: Transmission Line Theory and Microwave Applications", John wiley & Sons, 2005.

[21] Y. X Li, Y. S. Xie, H. W. Zhang, Y. L. Liu, Q. Y. Wen, W. W. Lin, " The strong non-reciprocity of metamaterial absorber: characteristic, interpretation and modelling", J Phys. D: Appl. Phys. 42 095408(2009)

[22] A. K. Azad, A. J. Taylor, E. Smirnova, J. F. O'Hara, " Characterization and analysis of terahertz metamaterials based on rectangular split-ring resonators", Appl. Phys. Lett. 92, 011119 (2008)

[23] L. Fu, H. Schweizer, H. Guo, N. Liu, H. Giessen, " Synthesis of transmission line models for metamaterial slabs at optical frequencies", Phys. Rev. B 78, 115110 (2008)

5

Metasurfaces for High Directivity Antenna Applications

Shah Nawaz Burokur, Abdelwaheb Ourir,
André de Lustrac and Riad Yahiaoui
Institut d'Electronique Fondamentale,
Univ. Paris-Sud, CNRS UMR 8622,
France

1. Introduction

There has been a lot of study published in literature on the improvement of the performances of microstrip patch antennas. Most of the solutions proposed in the past were to use an array of several antennas. The particular disadvantage of this method comes from the feeding of each antenna and also from the coupling between each element. Other interesting solutions have then been suggested: the first one (Jackson & Alexópoulos, 1985) was to make use of a superstrate of either high permittivity or permeability above the patch antenna and the second one proposed (Nakano et al., 2004), is to sandwich the antenna by dielectric layers of the same permittivity. A Left-Handed Medium (LHM) superstrate where both permittivity and permeability are simultaneously negative has also been suggested (Burokur et al., 2005). The numerical study of a patch antenna where a Left-Handed Medium (LHM) is placed above has been done and in this case a gain enhancement of about 3 dB has been observed. However, these solutions are all based on non-planar designs which are bulky for novel telecommunication systems requiring compact low-profile and environment friendly directive antennas.

To overcome the major problem of complex feeding systems in antenna arrays, the design of compact directive electromagnetic sources based on a single feeding point has become an important and interesting research field. Different interesting solutions based on this concept have been proposed. At first, resonant cavities in one-dimensional (1-D) dielectric photonic crystals have been used (Cheype et al., 2002). Afterwards, three dimensional (3-D) structures have been used, leading to better performances (Temelkuran et al., 2000). Another interesting solution proposed by Enoch *et al.* was to use the refractive properties of a low optical index material interface in order to achieve a directive emission (Enoch et al., 2002). The authors have shown how a simple stack of metallic grids can lead to ultra-refraction. Because the resulting metamaterial structure has an index of refraction, n, which is positive, but near zero, all of the rays emanating from a point source within such a slab of zero index material would refract, by Snell's Law, almost parallel to the normal of every radiating aperture. We shall note that these solutions are all also based on the use of a bulky 3-D material.

Otherwise, the most common method to reach directive emission is obviously based on the Fabry-Pérot reflex-cavity mechanism (Trentini, 1956). Such cavities have first been considered quite bulky too since a thickness of half of the working wavelength is required (Akalin et al., 2002). But recently, the introduction of composite metasurfaces has shown that the half wavelength thickness restriction in a Fabry-Pérot cavity can be judiciously avoided. For example, Feresidis *et al.* showed that a quarter wavelength thick Fabry-Pérot cavity can be designed by using Artificial Magnetic Conductor (AMC) surfaces introducing a zero degree reflection phase shift to incident waves (Feresidis et al., 2005). Assuming no losses and exactly 0° reflection phase, the surface is referred to as a Perfect Magnetic Conductor (PMC), which is the complementary of a Perfect Electric Conductor (PEC). The latter AMC surfaces have been first proposed in order to act as the so called High Impedance Surface (HIS) (Sievenpiper et al., 1999). This HIS is composed of metallic patches periodically organized on a dielectric substrate and shorted to the metallic ground plane with vias, appearing as "mushroom" structures. In a particular frequency band where reflection phase is comprised between -90° and +90°, this surface creates image currents and reflections in-phase with the emitting source instead of out of phase reflections as the case of conventional metallic ground plane. The HIS allows also the suppression of surface waves which travel on conventional ground plane. However, the HIS of Sievenpiper needs a non-planar fabrication process, which is not suitable for implementation in lots of microwave and millimetric circuits.

The reflex-cavity antenna proposed by Feresidis was composed of two planar AMC surfaces and a microstrip patch antenna acting as the primary (feeding) source. The first AMC surface was used as the feeding source's ground plane so as to replace the PEC surface and hence, to achieve a 0° reflection phase. The second one acted as a Partially Reflective Surface (PRS) with a reflection phase equal to 180°. This idea has then been pushed further by Zhou *et al.* (Zhou et al., 2005). By taking advantage of the dispersive characteristics of metamaterials, the authors designed a subwavelength cavity with a thickness smaller than a 10th of the wavelength. Compared to Feresidis, Zhou made use of a non-planar mushroom structure with a dipole acting as the feeding source.

In this chapter, using a novel composite metamaterial, made of both capacitive and inductive grids, we review our recent works in the fields of low-profile and high-gain metamaterial-based reflex-cavity type antennas. First, we will show how our group has lately further reduced the cavity thickness by $\lambda/30$ for applications to ultra-thin directive antennas by using a PEC surface as the source's ground plane and one subwavelength metamaterial-based composite surface as the PRS. We will also present how an optimization of the cavity has also been undertaken in order to reduce the thickness to $\lambda/60$ by using an AMC surface instead of the PEC ground plane and a metasurface as PRS. We will then present the modeling and characterization of resonant cavities for enhancing the directivity. Finally, a phase controlled metasurface will be proposed for applications to beam steerable and frequency reconfigurable cavity antennas. Numerical analyses using Finite Element Method (FEM) based software *HFSS* and CST's Transmission Line Modeling (TLM) solver *MICROSTRIPES* together with discussions on the fabrication process and the experimental results will be presented for the different cavities mentioned above.

2. Operating principle of the Fabry-Pérot reflex-cavity

A cavity antenna is formed by a feeding source placed between two reflecting surfaces as shown in Fig. 1. In this paper, different cavities based on the schematic model presented in Fig. 1 will be discussed and used. The cavity is composed of a PEC surface acting as a conventional ground plane for the feeding source and a metamaterial-based surface (metasurface) playing the role of a transmitting window known as a PRS. Following the earlier work of Trentini, a simple optical ray model can be used to describe the resonant cavity modes (Trentini, 1956). This model is used to theoretically predict the operating mode of a low-profile high-directivity metamaterial-based subwavelength reflex-cavity antenna. Let us consider the cavity presented in Fig. 1(a). It is formed by a feeding antenna placed between two reflectors separated by a distance h. Phase shifts are introduced by these two reflectors and also by the path length of the wave travelling inside the cavity. With the multiple reflections of the wave emitted by the antenna, a resonance is achieved when the reflected waves are in phase after one cavity roundtrip. The resonance condition, for waves propagating vertically, can then be written as:

$$h + t\sqrt{\varepsilon_r} = \left(\phi_{PRS} + \phi_r\right)\frac{\lambda}{4\pi} \pm N\frac{\lambda}{2} \qquad (1)$$

where ϕ_{PRS} is the reflection phase of the PRS reflector, ϕ_r is the reflection phase of the feeding source's ground plane, ε_r is the relative permittivity of the substrate supporting the primary source and t is its thickness. N is an integer qualifying the electromagnetic mode of the cavity. If the cavity and the substrate thicknesses t and h are fixed, the resonant wavelength is determined by the sum of the reflection phases $\phi_{PRS} + \phi_r$ for a fixed N. Conversely, for a given wavelength, the thickness h can be minimized by reducing the total phase shift $\phi_{PRS} + \phi_r$. The use of metasurfaces answers this purpose since they can exhibit an LC resonance. This resonance helps to have a reflection phase response varying from 180° to -180°, passing through 0° at the resonance frequency. By choosing a desired operating cavity frequency above the metasurface resonance where the reflection phase is negative, the sum $\phi_{PRS} + \phi_r$ can be very small leading to a very low cavity thickness. Since the reflector near the feeding antenna in Fig. 1(a) is composed of a PEC surface, then ϕ_r will be very close to 180°. On the other side, an AMC ground plane is used in Fig. 1(b) and in such case ϕ_r will show frequency dependent phase characteristics.

Therefore, taking advantage of the phase dispersive characteristics of metasurfaces, we will present several models of reflex-cavity antennas, each designed for a specific task. We will first present a λ/30 (1 mm @ 10 GHz) thick cavity antenna by using a PRS reflection phase value around -120°. This cavity antenna has a narrow beam profile in both E- and H-planes, producing a directivity of 160 (22 dBi). To further reduce the cavity thickness, we will emphasize on the use of two metasurfaces as illustrated in Fig. 1 (b), one as a PRS reflector and the other one as AMC ground plane of the primary source. The combination of these two metasurfaces, particularly the low phase values above their resonance, allows to design very low profile cavity antennas. For e.g., a λ/60 thickness has been achieved and the latter cavity presents a directivity of 78 (19 dBi).

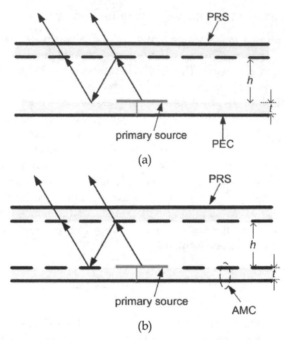

Fig. 1. Resonant cavity formed by a PEC ground plane and a metamaterial-based PRS (a) and, an AMC ground plane and a metamaterial-based PRS (b).

Since directivity depends strongly on the radiating aperture which is defined by the field distribution illuminating the PRS, we will present two ways on how we can manipulate the directivity of such reflex-cavity antennas. First, we will present the use of lateral PEC walls in the cavity antenna to form what we will refer to as metallic cavity. This method allows to enhance the directivity by 3 dB compared to the case where the cavity is open on the lateral sides. Also, the metallic cavity presents lower backward radiations due to the confinement of electromagnetic radiation, therefore increasing the front-to-back (FBR) ratio. Secondly, in order to optimize the field distribution illuminating the PRS, we will study the use of several primary sources inside the cavity. We will show how judiciously placing the different sources in the cavity helps to increase the directivity to more than 6 dB compared to single source fed cavity.

Finally, we will present beam steerable and frequency reconfigurable cavity antennas. For the beam steering, we will in a first step study a cavity where the PRS presents a locally variable phase. The latter PRS then acts as a phased array of micro-antennas, thus allowing to achieve beam steering. This concept has been pushed further by designing an electronically tunable metasurface via the incorporation of lumped elements (varactor diodes). This active metasurface can be used as PRS for two different tasks. Firstly, by applying different bias voltage along the PRS, a locally variable phase is obtained and is fully compatible for beam steering. On the other side, if we change the bias voltage of all the lumped elements similarly, then we can tune the operation frequency of the PRS so as to achieve a frequency reconfigurable reflex-cavity antenna.

3. Analysis of the planar metasurfaces

The cavity presented in Fig. 1 requires the application of a metamaterial-based surface. So in this section, we will design planar metamaterial-based surfaces for operation near 10 GHz.

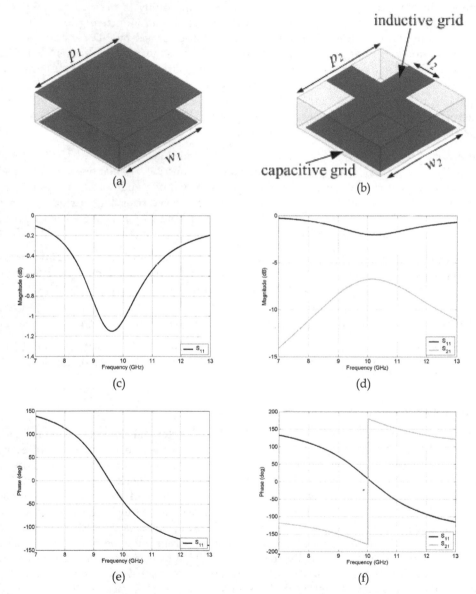

Fig. 2. Unit cell of AMC ground plane (a) and metamaterial-based PRS (b). Calculated reflection magnitude and phase of the AMC ground plane (c), (e) and reflection and transmission magnitudes and phases of the metamaterial-based PRS (d), (f).

The surface used by our group in order to achieve the AMC ground plane is made of a metamaterial composed of 2-D periodically subwavelength metallic square patches organized on one face of a dielectric substrate as illustrated in Fig. 2(a). The different dimensions of the patches are as follows: period p_1 = 4 mm and width w_1 = 3.8 mm. Another surface which we are going to use for the PRS of the cavity is made of a composite metamaterial consisting of simultaneously a capacitive and an inductive grid on the two faces of a dielectric substrate. The capacitive grid is also formed by 2-D periodic metallic patches (period p_2 = 4 mm and width w_2 = 3.6 mm) whereas the inductive grid is formed by a 2-D periodic mesh (line width l_2 = 1.2 mm) as shown in Fig. 2(b). Concerning the substrate, we have used the double copper cladded epoxy substrate of relative permittivity ε_r = 3.9, of tangential loss $\tan\delta$ = 0.0197 and having a thickness of 1.2 mm. The size of the different patterns has been chosen in order to minimize the phase of the reflection coefficient near 10 GHz while providing a sufficiently high reflectance (~90%).

The metasurfaces are analyzed numerically using the finite element software *HFSS* so as to present its characteristics in terms of reflection and transmission. Simulations are performed on a unit cell together with appropriate periodic boundary conditions. The results are presented in Fig. 2(c) and Fig. 2(d). As shown, the calculated resonance frequency of the AMC surface and PRS reflector is respectively 10.4 GHz and 9.7 GHz. At resonance, phase crosses 0° as illustrated in Fig. 2(e) and Fig. 2(f).

The composite metamaterial acts as a resonant filter which presents a reflection phase varying from 180° to –180°, depending on the frequency. This variation helps to be more flexible in designing thin cavities by choosing reflection phase values below 0°.

4. Metamaterial-based low-profile highly directive cavity antenna

In this section, we discuss about the design, implementation and characterization of low profile and highly directive cavity antennas. Two different models are presented; an AMC-PRS cavity and a PEC-PRS cavity.

4.1 AMC-PRS cavity antenna

The AMC-PRS cavity antenna is formed by the AMC reflector and the metasurface reflector used as PRS together with a patch antenna designed to operate near 10 GHz (Ourir et al., 2006a). The patch antenna of dimensions 6.8 × 7 mm² is placed on the AMC in the cavity as shown in Fig. 1(b). The reflectors used are those presented in Fig. 2. The different phases (simulated and measured) are used to estimate the thickness h of the AMC-PRS cavity as given by Eq. (1). Fig. 3(a) shows that h first decreases with increasing frequency of the first resonant mode (N = 0) to the point where a cavity zero thickness is reached at around 10.2 GHz. Then a jump in the mode occurs leading to an abrupt variation of h and the value decreases again for N = 1. A cavity thickness h = 0.5 mm is chosen for the cavity. The thickness h of the Fabry-Perot cavity formed by the two reflectors is adjusted mechanically. The lateral dimensions of the reflector plates are 17 × 17 cm². This thickness leads to a good matching of the cavity at 10.1 GHz (Fig. 3(b)) corresponding to the design of a $\lambda/60$ cavity. This frequency is in good agreement with the resonance frequency calculated from the optical ray model. The directive emission of the subwavelength cavity antenna at 10.1 GHz is illustrated from the calculated and measured E-plane (ϕ = 90°) and H-plane (ϕ = 0°) radiation patterns in Fig. 3(c) and 3(d).

Using the formulation proposed in (Temelkuran et al., 2000), the directivity of the cavity antenna is written as:

$$D = \frac{4\pi}{\theta_1 \theta_2} \tag{2}$$

where θ_1 and θ_2 are respectively the half-power widths for the E-plane and H-plane radiation patterns. The antenna directivity is then found to be equal to 78 (19 dB) for θ_1 = 22° and θ_2 = 24°.

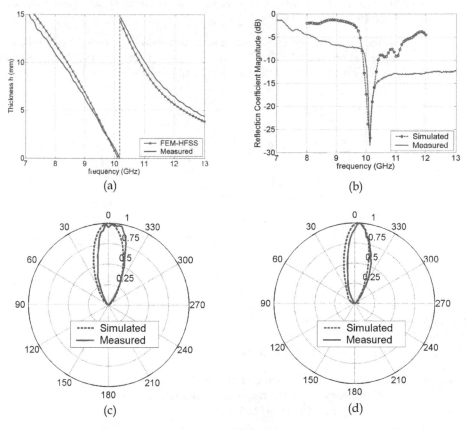

Fig. 3. (a) Evolution of the cavity thickness h versus frequency, this evolution being estimated from Eq. (1) by the calculated and measured reflection phases of the two reflectors used in the AMC-PRS cavity. (b) Calculated and measured matching of the cavity antenna. (c) E-plane (ϕ = 90°) radiation pattern at 10.1 GHz. (d) H-plane (ϕ = 0°) radiation pattern at 10.1 GHz.

4.2 PEC-PRS cavity antenna

In order to simplify the fabrication of the cavity antenna, another one using only one metamaterial-based surface reflector acting as the PRS and a PEC reflector (similar to the cavity shown in Fig. 1(a)) is designed (Ourir et al., 2006b). As we have seen from the reflection coefficients in Fig. 2(c) and 2(d), losses are maximum at the resonance frequency of the metamaterial-based surfaces. Thus using only one reflector has also the advantage of presenting lower losses. The PRS composed of simultaneously a capacitive and an inductive grid on the two faces of a dielectric substrate as presented in Fig. 2(b) has been designed for this purpose. Concerning the metallic patches of the capacitive grid, a period $p_2 = 5$ mm and a width $w_2 = 4.8$ mm are used. A line width $l_2 = 2.2$ mm is considered for the mesh of the inductive grid. This PRS having a resonance frequency of about 8 GHz presents a reflection phase close to -150° for frequencies higher than 10 GHz. The use of such a reflector in conjunction with a PEC leads also to a subwavelength cavity since the sum ($\phi_{PRS} + \phi_r$) is very close to zero between 9 GHz and 11 GHz.

A 1 mm ($\lambda/30$) thick cavity is designed with lateral dimensions of 10×10 cm² where the resonance is achieved at around 9.7 GHz. The antenna gain patterns in the E- and H-planes obtained from simulation and measurements are presented in Fig. 4(a) and 4(b).

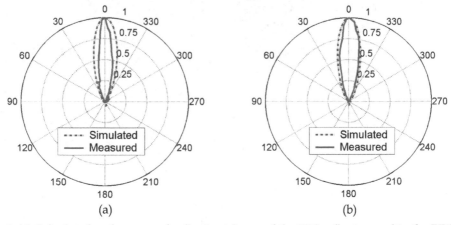

(a)   (b)

Fig. 4. (a) Calculated and measured reflection phases of the PRS reflector used in the PEC-PRS cavity. (b) Calculated and measured matching of the cavity antenna. (c) E-plane ($\phi = 90°$) radiation pattern at 9.7 GHz. (d) H-plane ($\phi = 0°$) radiation pattern at 9.7 GHz.

In this case, despite the use of only one metamaterial-based surface as the PRS and the use of smaller lateral dimensions than the two metamaterial-based cavity, the antenna directivity is found to be twice and equal to 160 (22 dB).

5. Directivity enhancement in Fabry-Pérot cavity antennas

This section deals with the enhancement of directivity in Fabry-Pérot cavity antennas. Two different approaches are presented to achieve higher performances in terms of directivity and beamwidths. In order to reach a higher directivity, a larger surface of the

PRS must be illuminated. Therefore, a better distribution and confinement of the electromagnetic energy must be produced in the cavity. For this purpose, two innovative solutions can be considered. The first one is to shield the cavity by four metallic walls and the second one is to feed the cavity by multiple primary sources. The two methods are detailed below.

5.1 Metallic cavity antenna

The cavity antenna proposed in this section was designed at 2.46 GHz for point to point radio communication links. The metallic cavity is composed of the feeding antenna's PEC ground plane and a metamaterial-based PRS as reflectors. Furthermore, four metallic walls are also fixed on the lateral sides so as to enhance the directivity of the cavity antenna while keeping low lateral dimensions (Burokur et al., 2009a).

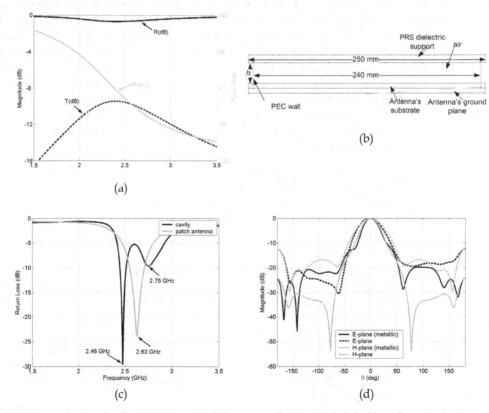

(a)

(b)

(c)

(d)

Fig. 5. (a) Calculated reflection phase (solid grey), reflection (solid dark) and transmission (dashed dark) magnitudes for the PRS reflector. (b) Schematic view of the metallic cavity antenna with h = 21.5 mm. (c) Return losses of the cavity antenna and the feeding patch antenna. (d) E- (ϕ = 90°) and H-plane (ϕ = 0°) radiation patterns at 2.46 GHz for the metallic and conventional cavity antennas.

The inductive and capacitive grids of the metasurface are printed on the faces of an 8 mm thick foam dielectric substrate (ε_r = 1.45, $\tan\delta$ = 0.0058). This thickness is sufficient enough to provide a relatively smooth slope of the phase response, hence rendering the metamaterial less sensitive to fabrication tolerances. The capacitive grid is formed by 2-D periodic metallic patches lattice (period p_2 = 20 mm and width w_2 = 18.8 mm) whereas the inductive grid is formed by a 2-D periodic mesh (line width l_2 = 6 mm). The size of the different patterns has been chosen in order to have the phase of the reflection coefficient below 0° near 2.46 GHz while providing a sufficiently high reflectance (~90%). The numerical results presented in Fig. 5(a) show firstly a resonance frequency of 2.38 GHz, i.e. where the phase crosses 0°. Secondly, we can also note a pass-band behavior where the transmission level is relatively low (about –9.5 dB). Finally this figure shows a reflection phase of –15° at 2.46 GHz.

The microstrip patch feeding source having dimensions 43 mm x 43 mm is designed on a similar foam dielectric substrate of thickness 5 mm. The surface of the inductive and capacitive grids forming the PRS has dimensions 200 mm × 200 mm, while the lateral dimensions of the dielectric board supporting the grids as well as that of the cavity have been increased to 250 mm × 250 mm. However the lateral metallic walls are separated by a distance of 240 mm, as illustrated by the side view of the cavity antenna in Fig. 5(b). So with a ϕ_{PRS} = –15°, the thickness of the cavity is found to be h = 21.5 mm (< $\lambda/5$).The simulated metallic cavity presents a return loss of 22.8 dB at 2.46 GHz [Fig. 5(c)]. A second resonance is observed at 2.75 GHz corresponding to the resonance of the feeding antenna. These two resonances are situated at each side of that of the feeding patch alone due to the coupling between the patch antenna and the FP cavity.

The calculated results [Fig. 5(d)] for the E- and H-plane radiation patterns show a directivity of 15.21 dB. Compared to a similar cavity without metallic walls, an enhancement of about 3 dB and lower secondary lobes are achieved. To reach this same directivity without metallic walls, we should have used a cavity with lateral dimensions close to 400 mm × 400 mm. Also, the metallic cavity presents very low backward radiations (–24.3 dB) due to the energy confinement by the lateral walls.

A prototype of the proposed cavity has been fabricated and measured (Fig. 6). However, the responses measured with h = 21.5 mm have not shown a resonance as expected at 2.46 GHz but at 2.49 GHz. This is due to the matching of the fabricated feeding patch antenna which does not occur at 2.63 GHz as in simulation. Moreover, the responses of the PRS may also present a shift in frequency which can be attributed to the manufacturing tolerances. A modification on the thickness of the cavity has then been undertaken in order to achieve as close as possible the calculated resonance frequency. Three other different thicknesses (h = 25 mm, h = 27.9 mm and h = 28.5 mm) have shown remarkable performances. The different results are summarized in Table 1.

As the thickness increases, the resonance of the cavity antenna tends to lower frequencies. For h = 25 mm, the measurements show a return loss of 11 dB and a directivity of 12.79 dB with secondary lobes reaching a level of –26.5 dB. For h = 28.5 mm, the return loss is enhanced to 21.5 dB at 2.405 GHz but the directivity falls to 12.4 dB. The best directivity (13.4 dB) is observed at 2.405 GHz for h = 27.9 mm with secondary lobes level of -22.7 dB.

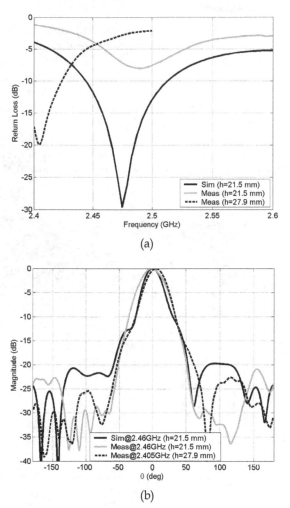

(a)

(b)

Fig. 6. (a) Measured return loss of the metallic cavity antenna. (b) Comparison between simulated and measured E-plane radiation patterns.

h (mm)	Resonant frequency (GHz)	Return loss (dB)	Directivity (dB)	Secondary lobes level (dB)
21.5 (sim)	2.46	29.5	15.3	-19.7
21.5 (meas)	2.49	8	12.36	-28.7
25 (meas)	2.46	11	12.79	-26.5
27.9 (meas)	2.405	20	13.4	-22.7
28.5 (meas)	2.4	21.5	12.4	-24.4

Table 1. Performances of the metallic cavity antenna.

5.2 Multisource-fed cavity antenna

As stated earlier, the second method to reach higher directivity is based on the use of multiple primary sources in the cavity. Therefore in this section, the cavities operating near 10 GHz are fed with a 2 x 2 microstrip patch array (Yahiaoui et al. 2009, Burokur et al., 2009b). The four patches with dimensions $W_p = L_p = 7.5$ mm are fed simultaneously via microstrip transmission lines acting as $\lambda/4$ impedance transformers and excited by a 50Ω SMA connector as shown in Fig. 7(a). The inter-element spacing a of the microstrip patch array feed plays an important role in the directivity of the cavity antenna. For this reason, the influence of this latter parameter is studied for a fixed cavity thickness h = 1.5 mm. The inter-element spacing a is varied from 0.5λ to 3λ. The return losses of the cavities are plotted in Fig. 7(b). We can note a very good matching (< –10 dB) around 9.25 GHz for the four different cases.

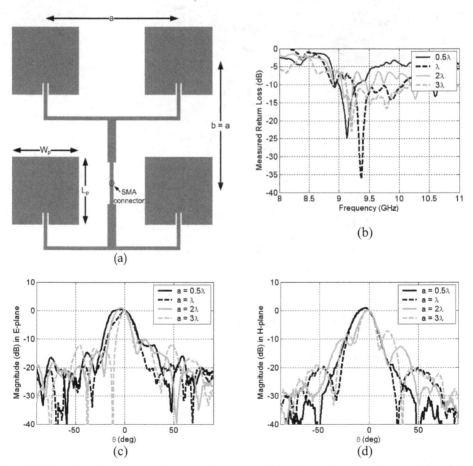

Fig. 7. (a) 2 × 2 patch array used as a multi-source. (b) Measured return losses of the cavities. (c)-(d) Measured E-plane and H-plane radiation patterns with a = 0.5λ, λ, 2λ and 3λ for a cavity thickness h = 1.5 mm.

The measured E- and H-plane radiation patterns of the cavity antennas are presented in Fig. 7(c) and 7(d). For $a = 0.5\lambda$, a measured directivity of 19 dB is obtained at 8.93 GHz. This value is very close to that of a cavity fed by a single source (see for e.g. Ourir et al., 2006a, 2006b). So, it is worth to note that conversely to classical antenna arrays, the directivity is not doubled each time that the number of sources is doubled. For $a = \lambda$, a measured directivity of 20.9 dB is noted at 9.07 GHz, showing clearly an enhancement of 1.9 dB with regard to the case $a = 0.5\lambda$. It is also very important to note that the sidelobes level of the patch array is considerably reduced when embedded in the cavity. This effect is highlighted in Table 2 where the performances of cavities for the different inter-element spacing are presented. 23.21 dB and 25.35 dB is respectively deduced from the measured planes for $a = 2\lambda$ and $a = 3\lambda$. When the case $a = 3\lambda$ is compared to $a = 0.5\lambda$, an increase of 6.35 dB is obtained for the directivity, which is comparable to an increase from a single patch element to a 2×2 patch array. The measured sidelobes level are higher (~ -8dB in the H-plane) for the case $a = 3\lambda$. However, this sidelobes level is still low compared to the sidelobes level of the source alone. It is well known that an inter-element spacing of an array higher than λ leads to high sidelobes level and also to the apparition of grating lobes.

The directivity D of the cavity antennas can be calculated using $D = 41253/(\theta_1 \times \theta_2)$ where θ_1 and θ_2 are respectively the half-power widths (in degrees) for the H-plane and E-plane patterns. The directivity values are given in Table 2 where we can observe that an increase in the inter-element spacing a in the cavity antenna gives rise to a higher directivity. This is because the radiation area at the surface of the source is bigger when a increases and therefore, a larger surface of the PRS is illuminated by the radiation source. This phenomenon is illustrated in Fig. 8 where the E-field distribution is plotted in a horizontal plane at two different locations z in the cavity antenna. $z = 0$ and $z = 1.5$ corresponds respectively to the plane of the radiating patch array source and to the thickness $h = 1.5$ mm at the inner surface of the PRS (location of the capacitive grid). This figure shows that the radiation area at the surface of the feed source in the case $a = 3\lambda$ is bigger than in the case $a = 0.5\lambda$ and therefore, a larger surface of the PRS is illuminated leading to a higher directivity. On the counter part, the side lobes level also increases.

a (mm)	Resonance frequency (GHz)	Maximum directivity (dB)	Secondary lobes level (dB)
0.5λ	9.13	19 @ 8.93 GHz	-12
λ	9.37	20.9 @ 9.07 GHz	-19
2λ	9.18	23.21 @ 8.94 GHz	-10
3λ	9.21	25.35 @ 8.96 GHz	-8

Table 2. Performances of the cavity antennas with $a = 0.5\lambda$, λ, 2λ and 3λ for a cavity thickness $h = 1.5$ mm.

6. Beam steering in Fabry-Pérot cavity antennas

In this section, we present the modeling and characterization of optimized resonant cavities for beam steering applications. Firstly, the design principle is presented for a passive cavity. The idea is then pushed further to achieve controllable beam steering by incorporating lumped elements in the metasurface reflector.

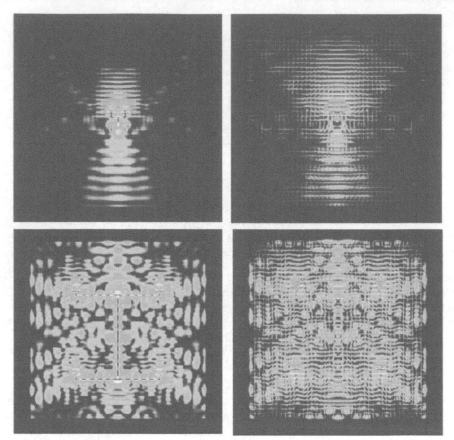

Fig. 8. E-field distribution in a horizontal plane in the cavity antenna for a = 0.5λ and a = 3λ.

6.1 Passive beam steering

Since the beam steering operation is presented in only one radiation plane, the metasurface used is composed of a 1-D array of copper strips etched on each face of a dielectric substrate as shown in Fig. 9(a).

We shall note that the gap spacing g in the capacitive grid plays a crucial role in determining the capacitance and therefore the resonance frequency of the metasurface. By changing g and keeping all the other geometric parameters unchanged, the capacitance of the metamaterial will also vary. As a consequence, the phases of the computed reflection coefficients vary. This behavior is illustrated by the numerical results shown in Fig. 9(b). We can note that the variation of g accounts for the shift of the resonance frequency. An increase in the value of g causes a decrease in the value of the capacitance created between two cells, and finally a shift of the resonance towards higher frequencies. At a particular frequency, the phase of the metasurface increases with an increase in the gap spacing. The study on the variation of g shows that it is possible to design a PRS with a continuous variation of the gap g, resulting in a local variation of the phase characteristics (Fig. 9(c)). If we consider each gap as a slot antenna, an analogy can then be made with an array of several antennas with a regular phase difference. The locally variable phase metasurface can then be applied for passive beam steering (Ourir et al. 2007a, 2009).

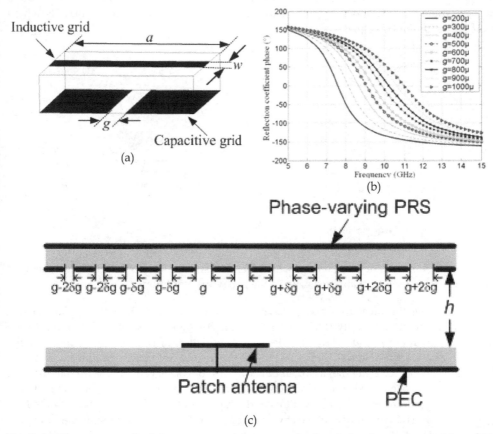

Fig. 9. (a) Elementary cell of the metamaterial composed of an inductive and capacitive grid, which is proposed for the PRS. (b) Reflection phase coefficient of the metasurface versus the gap width g. (c) Schematic view of the cavity composed of a PEC and a métasurface with a variable gap width.

To show the performances in terms of beam steering, several subwavelength cavities have been simulated and fabricated using the 1-D metasurface as PRS. The first one consists of the metamaterial PRS with the same gap spacing g = 400 µm between the metallic strips of the capacitive grid (δg = 0). This prototype will assure no deflection of the beam since it exists no phase variation of the metamaterial. The second and third ones are the prototypes incorporating respectively a variation of δg = 50 µm and δg = 100 µm along the positive x-direction. The cases where the variation δg is negative (180° rotation of the PRS around the z-axis) have also been considered. Note that here the resonance frequency of the central region of the metamaterial corresponds to that of the PRS without gap spacing variation (g = 400 μm and δg = 0), i.e. 8.7 GHz as shown in Fig. 9(b). The resonance frequency of the cavity is found to be ~10.5 GHz for the three prototypes as shown in Fig. 10(a). Best matching is observed when the metallic gap of the PRS capacitive grid increases. However, the resonance frequency remains the same for the three configurations since it depends on the gap spacing of the central region of the PRS, which is the same for the three prototypes.

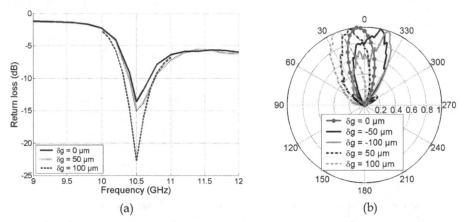

Fig. 10. (a) Return loss of the antennas with different variation of gap width. (b) Measured gain patterns of the cavity antennas versus the gap width variation.

Fig. 10(b) shows the measured gain patterns of the antenna in the E (ϕ = 90°) plane at 10.5 GHz for an optimized cavity thickness h = 1 mm. For δg = 0, the beam is normal to the plane of the antenna and shows no deflection, which confirms our prediction on the constant phase metamaterial. However, in the case of a regular variation of 50 μm, a deflection of the antenna beam of about 10° can be observed either in the forward (clockwise) or backward (anti-clockwise) direction depending if δg is respectively negative or positive. Similar observations and a higher deflection of ±20° can be noted for δg = ±100 μm. The directivity of the cavity antenna can be calculated using the following expression: D = 41253/($\theta_1 \times \theta_2$) where θ_1 and θ_2 are respectively the half-power widths (in degrees) for the H-plane and E-plane patterns. In this case, the directivity is found to be approximately equal to 14.8 dB.

6.2 Active beam steering

The cavity antenna proposed in this section includes the use of lumped elements such as varactor diodes so as to be able to control electronically the phase of the metasurface. As a preliminary step in the design of such cavities, we present firstly the design of the active metasurface.

6.2.1 Electronically controlled metasurface

The metasurface used in this section is based on the same principle as the one illustrated in Fig. 9. But, instead of applying a linear variation of the gap spacing g in order to create a locally variable phase, we now use active components to make the phase of the metasurface shift in frequency. Varactor diodes having a capacitance value ranging from 0.5 pF to 1.0 pF are thus incorporated into the capacitive grid between two adjacent metallic strips (Fig. 11(a)) and depending on the applied bias voltage, the phase of the metasurface varies with frequency (Ourir et al. 2007b). The variable capacitive grid of the tunable phase PRS used for an operating frequency around 8 GHz consists of a lattice of metallic strips with varactor diodes connected each 6 mm (s = 6 mm) between two adjacent strips. The width of the strips and the spacing between two strips of the capacitive grid is respectively w = 1 mm and g = 2 mm.

Concerning the inductive grid, the width of the strips and the spacing between two strips are respectively w_1 = 2 mm and g_1 = 4 mm (Fig. 11(b)). Note that the inductive grid is not made tunable. RF chokes are also used in the microstrip circuit in order to prevent high frequency signals going to the DC bias system. Potentiometers are implemented in the structure to create a voltage divider circuit so as to be able to bias locally the varactors. The capacitance in each row can then be adjusted according to the bias voltage applied. This capacitance can also be varied from one row to another by the use of the voltage dividers on the prototype. By changing the bias voltage of the varactors of the PRS similarly, the capacitance of the metamaterial will also vary. As a consequence, the reflection and the transmission coefficients also vary. This behavior is illustrated by the measurement results of the reflection coefficient magnitude and phase shown in Fig. 11(c) and 11(d) respectively. These curves are obtained when the same bias voltage is applied to the different rows of varactors along the PRS. The measurements are performed in an anechoic chamber using two horn antennas working in the [2 GHz – 18 GHz] frequency band and an 8722ES network analyzer. From Fig. 11(c), we can note that the variation of the bias voltage accounts for the shift of the resonance frequency of the PRS, i.e. the frequency where the phase crosses 0°. An increase in the bias voltage leads to a decrease in the value of the capacitance of the metamaterial, and finally to a shift of the resonance towards higher frequencies. At a particular frequency the phase of the PRS increases with an increase in the bias voltage. This phase shift is very important since it will help to tune the resonance frequency of the cavity antenna and also to control the radiated beam direction of the antenna.

6.2.2 Active beam steering

Instead of applying a uniform variation in the periodicity of the cells composing the capacitive grid so as to create a locally variable phase as in section 6.1, we now use the electronically controlled metasurface as PRS (Ourir et al. 2009). The active components biased differently make the phase of the PRS shifts in frequency locally. As illustrated by the varactors bias system shown in Fig. 12(a), the proposed PRS is now divided into different regions, where each of them has a specific bias voltage bias. We shall note that here the resonance frequency of the cavity is imposed by the resonance frequency of the central region just above the feeding source corresponding to the bias voltage $V_4=V_1+3\delta V$. The bias voltage is thus increased uniformly with a step δV when moving from the left to the right of the metamaterial-based PRS by the use of the potentiometers. This action creates a regular variation of the phase along the PRS.

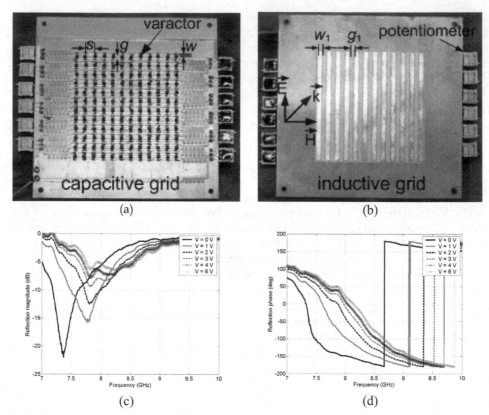

(a)   (b)

(c)   (d)

Fig. 11. (a) Electronically phase-varying metasurface. (a) Capacitive grid incorporating varactors and voltage dividers. (b) Inductive grid. (c) Measured magnitude and (d) measured phase of the reflection coefficient versus bias voltage of the varicaps.

The first configuration studied here is the antenna cavity based on the metamaterial PRS with the same null bias voltage for all the varactors. This configuration will assure no deflection of the beam since it exists no phase variation of the metamaterial. The second and third configurations are prototypes incorporating respectively a variation of $\delta V = 0.2$ V and $\delta V = 0.3$ V along the positive x-direction. The cases where the variation δV is negative (180° rotation of the PRS around the z-axis) have also been considered.

Fig. 12(b) shows the gain patterns of the antenna in the E-plane ($\phi = 90°$) at 7.9 GHz for the optimized cavity. For $\delta V = 0$ V, the beam is normal to the plane of the antenna and shows no deflection, which confirms our prediction on the constant phase metamaterial. However, in the case of a regular variation of $\delta V = 0.2$ V, a deflection of the antenna beam of about 7° can be observed either in the forward or backward direction depending if δV is respectively negative or positive. Similar observations and a higher deflection can be noted for respectively $\delta V = 0.3$ V and $\delta V = -0.3$ V. This figure illustrates very clearly the control of the radiation pattern of the antenna by the bias voltage of the varactors. The direction of the radiation beam depends of the direction of the variation of the bias of the varactors. If we inverse the sign of δV, the sign of the deviation changes also. This demonstration opens the door to the realization of very simple electronically beam steering ultra-compact antennas based on active metamaterials.

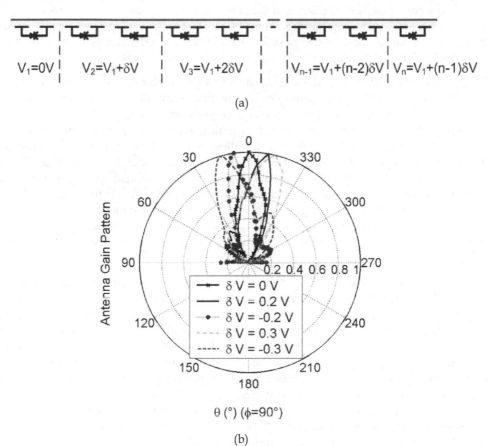

(a)

(b)

θ (°) (φ=90°)

Fig. 12. (a) Variation of the bias voltage of the varactors along the phase varying PRS. (b) Measured gain patterns in the E-plane (φ = 90°) at 7.9 GHz for δV = 0 V, δV = 0.2 V and δV = 0.3 V. The steering of the antenna's radiated beam can be clearly observed with a positive steering angle for positive bias and negative one for a negative bias.

7. Frequency agile Fabry-Pérot cavity antennas

Conversely to beam steerable cavity antennas, we do not need a locally phase-varying PRS for frequency agility applications. What we seek is the ability to change the resonance frequency of the PRS and this is possible by changing simultaneously and in the same manner the capacitance value of the varactor diodes. Here, we show that a tunable metasurface associated to an array of wideband sources in a Fabry-Pérot cavity leads to a reconfigurable directive emission on a wide frequency range (Burokur et al. 2010, 2011). A similar electronically controlled PRS as the one shown in Fig. 11(a) is designed to operate near 2 GHz in base station antennas for mobile phone communication systems. The primary source of the cavity is a wideband microstrip patch antenna designed to cover 1.8 GHz – 2.7 GHz frequency range and therefore to illuminate the PRS at any frequency within this range. This patch antenna is electromagnetically coupled to an L-probe which itself is connected to a coaxial connector. Simulations have shown a good matching (return loss < 10 dB) from 1.8 GHz to 2.7 GHz.

To demonstrate experimentally the mechanism for reconfigurable directive emissions from a metamaterial-based FP cavity, a prototype having dimensions 400*400 mm² (approximately $3\lambda*3\lambda$) has been fabricated and tested. As it has been shown in section 5.2, the directivity is drastically enhanced when a cavity is fed by judiciously spaced multiple sources since a larger surface of the PRS is illuminated, and therefore the size of the effective radiating aperture of the cavity antenna is increased. Four elementary sources constituting a 2 x 2 wideband patch array are used as primary source; the inter-element spacing between the different sources being 200 mm. Fig. 13(a) and 13(b) shows respectively the photography of the prototype and the capacitive grid of the electronically tunable metasurface In order to experimentally estimate directivity and gain of the cavity's radiated beam, direct far field measurements are performed using a SATIMO STARLAB and the characteristics are shown in Fig. 14.

When capacitance of the metasurface reflector is changed by varying bias voltage of varactor diodes, the frequency of maximum gain is tuned as clearly shown in the different diagrams of Fig. 14. When 0 V is applied, maximum gain is observed at 1.9 GHz corresponding approximately to the simulated case with C = 6.5 pF. When DC bias voltage is increased, the capacitance value is decreased, resulting in an increase of maximum gain frequency. For 24 V, maximum gain occurs at 2.31 GHz, corresponding to lowest capacitance value. To gain more insight in the electromagnetic properties of the metamaterial-based Fabry-Pérot cavity, intensity maps of scanned far field versus frequency and elevation angle θ, in E-plane are presented. The emission frequency represented by the red spot varies from 1.9 GHz to 2.31 GHz from 0 V to 24 V as shown in Figs. 14(a), 14(c), 14(e) and 14(g). These figures demonstrate clearly the frequency reconfigurability property of the cavity. We shall also note that for each frequency the spot is situated at an elevation angle of 0°, indicating a radiated beam normal to the cavity metasurface reflector. Figs. 14(b), 14(d), 14(f) and 14(h) show radiation patterns in E- and H-planes at respectively 1.9 GHz, 2.02 GHz, 2.16 GHz and 2.31 GHz corresponding to maximum gain frequency for 0 V, 5 V, 12 V and 24 V. The tuning range of maximum gain frequency results in an effective operation bandwidth close to 20%. A wide frequency bandwidth is achieved due to the cavity thickness fixed in this particular case. With h = 15 mm, reflection phase values around 0° are needed in the 1.85 GHz – 2.25 GHz frequency band. A lower h would lead to phase values approaching -180° and the possible frequency bandwidth from the capacitance tuning range would be narrow. Actually, a high directivity (approximately 18 dBi) is obtained experimentally due to the large lateral dimensions of the fabricated cavity and also to the use of four elementary sources instead of only one where only 14 dBi is obtained.

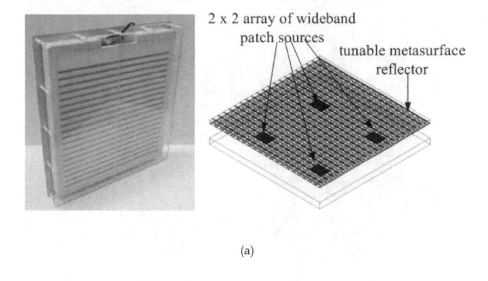

2 x 2 array of wideband patch sources

tunable metasurface reflector

(a)

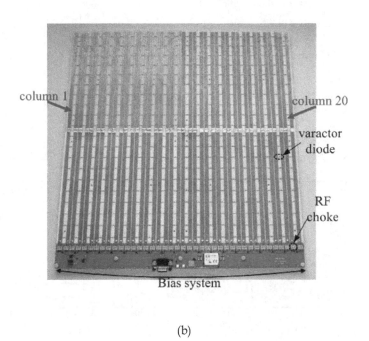

column 1

column 20

varactor diode

RF choke

Bias system

(b)

Fig. 13. (a) Photography and perspective view of the cavity antenna. (b) Electronically tunable metasurface reflector incorporating varactor diodes, RF chokes and bias system.

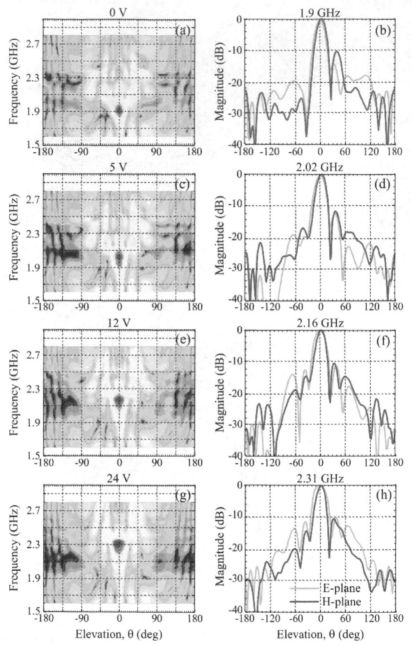

Fig. 14. Far field intensity maps versus frequency and elevation angle in E-plane and measured radiation patterns in E- and H-planes at maximum gain frequency for different bias voltage applied : (a)-(b) 0 V – 1.9 GHz, (c)-(d) 5 V – 2.02 GHz, (e)-(f) 12 V – 2.16 GHz, and (g)-(h) 24 V – 2.31 GHz.

8. Conclusion

To conclude, we have presented various aspects of reflex-cavity antennas: low-profile, high gain, beam steering and frequency agility. For each aspect, numerical calculations together with measurements have been presented. The development of these works has enabled to promote the interesting characteristics of metamaterial-based surfaces. Variable phase metasurfaces compared to conventional PEC and AMC surfaces have also shown their usefulness in reconfigurability applications. Further studies are actually performed to transpose the reflex-cavity antenna concept to industrial applications in various domains such as telecommunications, aeronautical, transport and housing.

9. Acknowledgements

The authors are very grateful to the French National Research Agency (ANR) for the financial support of the METABIP Project. These works have also been made possible by the partial financial support of the Eureka TELEMAC project. We would like also to thank our partners P. Ratajczak and J.-P. Daniel for the fabrication and characterization of antenna prototypes.

10. References

Akalin, T., Danglot, J., Vanbesien, O. & Lippens, D. (2002). A highly directive dipole antenna embedded in a Fabry–Perot-type cavity. *IEEE Microw. Wireless Component Lett.*, Vol.12, No.2, (February 2002), pp. 48-50, ISSN 1531-1309.

Burokur, S.N., Latrach, M. & Toutain, S. (2005). Theoretical investigation of a circular patch antenna in the presence of a Left-Handed Medium. *IEEE Antennas Wireless Propag. Lett.*, Vol.4, (June 2005), pp. 183-186, ISSN 1536-1225.

Burokur, S.N., Ourir, A., Daniel, J.-P., Ratajczak, P. & de Lustrac, A. (2009a). Highly directive ISM band cavity antenna using a bi-layered metasurface reflector. *Microwave Opt. Technol. Lett.*, Vol.51, No.6, (June 2009), pp. 1393-1396, ISSN 0895-2477.

Burokur, S.N., Yahiaoui, R. & de Lustrac, A. (2009b). Subwavelength metamaterial-based resonant cavities fed by multiple sources for high directivity. *Microwave Opt. Technol. Lett.*, Vol.51, No.8, (August 2009), pp. 1883-1888, ISSN 0895-2477.

Burokur, S.N., Daniel, J.-P., Ratajczak, P. & de Lustrac, A. (2010). Tunable bilayered metasurface for frequency reconfigurable directive emissions. *Appl. Phys. Lett.*, Vol.97, No.6, (August 2010), 064101, ISSN 0003-6951.

Burokur, S.N., Daniel, J.-P., Ratajczak, P. & de Lustrac, A. (2011). Low-profile frequency agile directive antenna based on an active metasurface. *Microwave Opt. Technol. Lett.*, Vol.53, No.10, (October 2011), pp. 2291-2295, ISSN 0895-2477.

Cheype, C., Serier, C., Thèvenot, M., Monédière, T., Reinex, A. & Jecko, B. (2002). An electromagnetic bandgap resonator antenna. *IEEE Trans. Antennas Propag.*, Vol.50, No.9, (September 2002), pp. 1285-1290, ISSN 0018-926X.

Enoch, S., Tayeb, G., Sabouroux, P., Guérin, N. & Vincent, P. (2002). A metamaterial for directive emission. *Phys. Rev. Lett.*, Vol.89, No.21, (November 2002), 213902, ISSN 0031-9007.

Feresidis, A.P., Goussetis, G., Wang, S. & Vardaxoglou, J.C. (2005). Artificial Magnetic Conductor Surfaces and their application to low-profile high-gain planar antennas. *IEEE Trans. Antennas Propag.*, Vol.53, No.1, (January 2005), pp. 209-215, ISSN 0018-926X.

Jackson, D.R. & Alexópoulos, N.G. (1985). Gain enhancement methods for printed circuit antennas. *IEEE Trans. Antennas Propag.*, Vol.AP-33, No.9, (September 1985), pp. 976-987, ISSN 0018-926X.

Nakano, H., Ikeda, M., Hitosugi, K. & Yamauchi, J. (2004). A spiral antenna sandwiched by dielectric layers. *IEEE Trans. Antennas Propag.*, Vol.52, No.6, (June 2004), pp. 1417-1423, ISSN 0018-926X.

Ourir, A., de Lustrac, A. & Lourtioz, J.-M. (2006a). All-metamaterial-based subwavelength cavities ($\lambda/60$) for ultrathin directive antennas. *Appl. Phys. Lett.*, Vol.88, No.8, (February 2006), 084103, ISSN 0003-6951.

Ourir, A., de Lustrac, A. & Lourtioz, J.-M. (2006b). Optimization of metamaterial based subwavelength cavities for ultracompact directive antennas. *Microwave Opt. Technol. Lett.*, Vol.48, No.12, (December 2006), pp. 2573-2577, ISSN 0895-2477.

Ourir, A., Burokur, S.N. & de Lustrac, A. (2007a). Phase-varying metamaterial for compact steerable directive antennas. *Electron. Lett.*, Vol.43, No.9, (April 2007), pp. 493-494, ISSN 0013-5194.

Ourir, A., Burokur, S.N. & de Lustrac, A. (2007b). Electronically reconfigurable metamaterial for compact directive cavity antennas. *Electron. Lett.*, Vol.43, No.13, (June 2007), pp. 698-700, ISSN 0013-5194.

Ourir, A., Burokur, S.N., Yahiaoui, R. & de Lustrac, A. (2009). Directive metamaterial-based subwavelength resonant cavity antennas – Applications for beam steering. *C. R. Physique*, Vol.10, No.5, (June 2009), pp. 414-422, ISSN 1631-0705.

Sievenpiper, D., Zhang, L., Broas, R.F.J., Alexópoulos, N.G. & Yablonovitch, E. (1999). High-Impedance Electromagnetic Surfaces with a forbidden frequency band. *IEEE Trans. Microw. Theory Tech.*, Vol.47, No.11, (November 1999), pp. 2059-2074, ISSN 0018-9480.

Temelkuran, B., Bayindir, M., Ozbay, E., Biswas, R., Sigalas, M.M., Tuttle, G. & Ho, K.M. (2000). Photonic crystal-based resonant antenna with a very high directivity. *J. Appl. Phys.*, Vol.87, No.1, (January 2000), pp. 603-605, ISSN 0021-8979.

Trentini, G.V. (1956). Partially reflecting sheet arrays. *IRE Trans. Antennas Propag.*, Vol.4, No.4, (October 1956), pp. 666-671, ISSN 0096-1973.

Yahiaoui, R., Burokur, S.N. & de Lustrac, A. (2009). Enhanced directivity of ultra-thin metamaterial-based cavity antenna fed by multisource. *Electron. Lett.*, Vol.45, No.16, (July 2009), pp. 814-816, ISSN 0013-5194.

Zhou, L., Li, H., Qin, Y., Wei, Z. & Chan, C.T. (2005). Directive emissions from subwavelength metamaterial-based cavities. *Appl. Phys. Lett.*, Vol.86, No.10, (March 2005), 101101, ISSN 0003-6951.

6

Metamaterial-Based Compact Filter Design

Merih Palandöken
Berlin Institute of Technology
Germany

1. Introduction

Emerging requirements of increasingly complex wireless systems necessitate novel design methods of wireless components to be developed for the fulfillment of many performance criteria simultaneously. One of these wireless components to be enhanced for high data rate transmission systems and matched with the new challenges in new generation communication systems is the microwave filter. This chapter is mainly dealing with the novel microwave filter design methods based on artificial materials. How to design electrically small resonators and to couple each of these resonators with the successive resonators in a periodic/aperiodic manner in addition to the feeding line for the bandpass and bandstop filter designs are highlighted throughout this chapter.

In this chapter, because the current trend in filter design is the filter miniaturization for more compact wireless systems, basic approaches in electrically small filter design based on artificial metamaterials are explained. The compact resonators result the signal suppression level of the incoming signal in bandstop filter designs and signal selectivity of the transmitted signal in bandpass filter designs to be higher as a performance enhancement. It is due to the target geometrical compactness, resulting into the possibility of cascading more electrically small resonator cells in a restricted area. Because these performance parameters are highly related with Q factor of each individual resonator, the eigenmodes of periodically loaded negative permittivity and negative permeability resonators are analytically calculated in Section 2. The passband frequencies and Q factor of the filter to be designed can be engineered with the geometrical and topological parameters of the resonators, which is the fundamental idea of artificial material based filter designs. This analytical calculation points out the effect of electromagnetic material parameters of negative permittivity and permeability materials on filter performance. In Section 3 and Section 4, two compact filter designs are proposed and explained along with the numerical results as metamaterial based filter examples. One of these designs is the fractal resonator based bandstop filter. The other design is the compact bandpass filter based on thin wire loaded spiral resonator cells. These are quite good examples of compact filter designs with negative permeability and left-handed metamaterial unit cells, respectively. These numerically calculated filters points out the compactness and performance enhancement of novel designs in comparison to the conventional filter designs in a restricted volume.

2. Theoretical analysis

The motivating principle of theoretical research on LHM is first introduced by Veselago with his theoretical paper in 1968 [Veselago, 1968]. He considered electromagnetic wave propagation through a homogenous isotropic electromagnetic material in which both permittivity and permeability were assumed to have negative real values. Because the direction of the Poynting vector of a monochromatic plane wave is opposite to that of its phase velocity in such a material, he referred to this medium as left-handed medium. This material property makes in turn these materials to support backward-wave propagation due to its negative refractive index. The natural inexistence of such exotic materials had unfortunately led the motivating ideas of Veselago on negative refraction, its various electromagnetic and optical consequences to receive little attention in the scientific community [Engheta & Ziolkowski, 2006; Caloz & Itoh, 2005; Eleftheriades & Balmain, 2005]. However, the work of Pendry on the electromagnetic engineering of magnetic permeability and electric permittivity of the materials with electrically small metallic inclusions has made Veselago's ideas realisable [Pendry et al.,1999; Pendry et al., 1996]. In 2000, Smith inspired by the work of Pendry and constructed a composite LH medium in the microwave regime by arranging the periodic arrays of small metallic wires and SRRs [Smith & Kroll, 2000]. He demonstrated the anomalous refraction at the interface of this LH medium with the air, which is the result of negative refraction in this artificial material. The effective electromagnetic parameters were also retrieved experimentally and numerically from the transmission and reflection data to prove the negative refractive index [Smith et al., 2005 ; Alexopoulos et al., 2007; Chen et al., 2004 ; Smith et al., 2000]. There have now been several theoretical and experimental studies that have been reported confirming negative refractive index. There are some engineering applications derived from this concept such as phase compensation and electrically small resonators [Engheta, 2002], negative angles of refraction [Kong et al., 2002; Kolinko & Smith, 2003; Ziolkowski, 2003], sub-wavelength waveguides with lateral dimensions below diffraction limits [Alu & Engheta, 2003,2004], enhanced focusing [Grbic & Eleftheriades,2004], backward wave antennas [Grbic & Eleftheriades,2002; Caloz & Itoh, 2005], enhanced electrically small antennas [Ziolkowski & Kipple, 2003] and compact microwave filters [Marques et al., 2008].

The main principle of negative refractive index in LHMs can be deduced quite easily by the calculation of TE and TM wave impedances of waveguide modes. The TE_{mn} and TM_{mn} wave impedances are calculated in a general form as

$$Z_{TEmn} = \frac{\omega\mu}{k_{zmn}} = -\frac{\omega\left|\mu_{LHM}\right|}{k_{zmn}}$$

$$Z_{TMmn} = \frac{k_{zmn}}{\omega\varepsilon} = -\frac{k_{zmn}}{\omega\left|\varepsilon_{LHM}\right|}$$

(1)

with the longitudinal wave number, k_{zmn} at the operation frequency of ω with the magnetic permeability, μ and electric permittivity, ε of the waveguide filling material. Because the wave impedance of a passive microwave component is always positive, the phase constant has to be correspondingly negative in the impedance formulation due to the negative permittivity and permeability. As a result, the effective refractive index is also negative. This is an alternative explanation to understand the underlying reasoning why the refractive index has to be negative in the artificial materials with negative permittivity and permeability. Due to the exploitation of electrically small resonators in the resonator design, compact resonators can be cascaded to increase the frequency selectivity and stop-band rejection level in a restricted volume. The accompanying degradation of Q factor due to the periodic arrangement of intercoupled cells can be compensated by the implementation of complementary resonators, in which virtual magnetic currents are the excited resonant sources instead of electrical currents flowing through the lossy metal [Marques et al., 2008].

In this chapter, the subwavelength resonance feature of periodically arranged negative permittivity and permeability materials in a rectangular waveguide is explained by the calculation of eigenmode equation for TE modes. The eigenmode calculation can be similarly done for TM modes by replacing TE wave impedance with TM wave impedance in ABCD matrix formulation. The periodically loaded waveguide model is shown in Fig. 1.

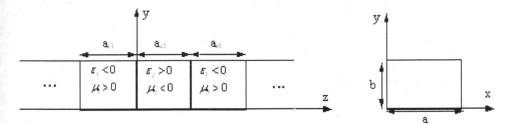

Fig. 1. Waveguide model of periodically arranged negative permeability and permittivity materials

ABCD matrix of negative permittivity material of length a_{z1} is formulated with TE-wave impedance, Z_{H1} and phase constant, k_{z1} as

$$M_{ABCD\_negeps} = \begin{bmatrix} \cos(k_{z1}a_{z1}) & jZ_{H1}\sin(k_{z1}a_{z1}) \\ j\dfrac{\sin(k_{z1}a_{z1})}{Z_{H1}} & \cos(k_{z1}a_{z1}) \end{bmatrix}$$

$$k_{z1} = \sqrt{\omega^2\mu_1\varepsilon_1 - (\frac{m\pi}{a})^2 - (\frac{n\pi}{b})^2} \qquad (2.a)$$

$$Z_{H1} = \frac{\omega\mu_1}{k_{z1}}$$

where a and b are the waveguide side lengths in x and y directions, respectively.

In the same manner, ABCD matrix of negative permeability material of length a_{z2} is formulated with TE-wave impedance, Z_{H2} and phase constant, k_{z2} as

$$M_{ABCD\_negmu} = \begin{bmatrix} \cos(k_{z2}a_{z2}) & jZ_{H2}\sin(k_{z2}a_{z2}) \\ j\dfrac{\sin(k_{z2}a_{z2})}{Z_{H2}} & \cos(k_{z2}a_{z2}) \end{bmatrix}$$

$$k_{z2} = \sqrt{\omega^2 \mu_2 \varepsilon_2 - (\frac{m\pi}{a})^2 - (\frac{n\pi}{b})^2}$$

$$Z_{H2} = \frac{\omega \mu_2}{k_{z2}} \qquad (2.b)$$

Thus, ABCD matrix of one unit cell of length, $a_{z1}+a_{z2}$, consisting of the negative permeability and permittivity materials is calculated by cascading ABCD matrices of the respective materials as

$$M_{ABCD\_cell} = \begin{bmatrix} \cos(k_{z2}\frac{a_{z2}}{2}) & jZ_{H2}\sin(k_{z2}\frac{a_{z2}}{2}) \\ j\dfrac{\sin(k_{z2}\frac{a_{z2}}{2})}{Z_{H2}} & \cos(k_{z2}\frac{a_{z2}}{2}) \end{bmatrix} \begin{bmatrix} \cos(k_{z1}a_{z1}) & jZ_{H1}\sin(k_{z1}a_{z1}) \\ j\dfrac{\sin(k_{z1}a_{z1})}{Z_{H1}} & \cos(k_{z1}a_{z1}) \end{bmatrix}$$

$$\begin{bmatrix} \cos(k_{z2}\frac{a_{z2}}{2}) & jZ_{H2}\sin(k_{z2}\frac{a_{z2}}{2}) \\ j\dfrac{\sin(k_{z2}\frac{a_{z2}}{2})}{Z_{H2}} & \cos(k_{z2}\frac{a_{z2}}{2}) \end{bmatrix}$$

$$M_{ABCD\_cell} = \begin{bmatrix} A & B \\ C & D \end{bmatrix} \qquad (2.c)$$

where the matrix elements are calculated as in (2.d).

$$A = \cos(k_{z1}a_{z1})\cos(k_{z2}a_{z2}) - \frac{1}{2}\left(\frac{Z_{H2}}{Z_{H1}} + \frac{Z_{H1}}{Z_{H2}}\right)\sin(k_{z1}a_{z1})\sin(k_{z2}a_{z2})$$

$$B = jZ_{H2}\cos(k_{z1}a_{z1})\sin(k_{z2}a_{z2}) + j\frac{\sin(k_{z1}a_{z1})}{2Z_{H1}}\left((Z_{H1}^2 - Z_{H2}^2) + (Z_{H1}^2 + Z_{H2}^2)\cos(k_{z2}a_{z2})\right)$$

$$C = j\frac{1}{Z_{H2}}\cos(k_{z1}a_{z1})\sin(k_{z2}a_{z2}) + j\frac{\sin(k_{z1}a_{z1})}{2Z_{H1}Z_{H2}^2}\left((-Z_{H1}^2 + Z_{H2}^2) + (Z_{H2}^2 + Z_{H1}^2)\cos(k_{z2}a_{z2})\right) \qquad (2.d)$$

$$D = \cos(k_{z1}a_{z1})\cos(k_{z2}a_{z2}) - \frac{1}{2}\left(\frac{Z_{H2}}{Z_{H1}} + \frac{Z_{H1}}{Z_{H2}}\right)\sin(k_{z1}a_{z1})\sin(k_{z2}a_{z2})$$

Thus, the dispersion relation of the periodic array of negative permeability and permittivity materials is calculated from ABCD parameters to determine the complex propagation constant, γ_{zeff} as

$$\cosh(\gamma_{zeff}(a_{z1} + a_{z2})) = \cos(k_{z1}a_{z1})\cos(k_{z2}a_{z2}) - \frac{1}{2}\left(\frac{Z_{H1}}{Z_{H2}} + \frac{Z_{H2}}{Z_{H1}}\right)\sin(k_{z1}a_{z1})\sin(k_{z2}a_{z2})$$

$$(2.e)$$

This dispersion relation can be formulated in an alternative form for the case of ideal lossless material parameters as

$$\cosh(\gamma_{zeff}(a_{z1}+a_{z2}))=\cosh(\alpha_{z1}a_{z1})\cosh(\alpha_{z2}a_{z2})-\frac{1}{2}\left(\frac{Z_1}{Z_2}+\frac{Z_2}{Z_1}\right)\sinh(\alpha_{z1}a_{z1})\sinh(\alpha_{z2}a_{z2}) \quad (3.a)$$

with the wave reactances, $Z_{1,2}$ and attenuation constants, $\alpha_{z1,2}$ of the negative permittivity and permeability media, respectively. The wave reactances and attenuation constants are calculated from (2.a) and (2.b) in the form of

$$Z_{H1}=\frac{\omega\mu_1}{-j\sqrt{\omega^2\mu_1|\varepsilon_1|+(\frac{m\pi}{a})^2+(\frac{n\pi}{b})^2}}=jZ_1$$

$$Z_{H2}=\frac{\omega|\mu_2|}{j\sqrt{\omega^2|\mu_2|\varepsilon_2+(\frac{m\pi}{a})^2+(\frac{n\pi}{b})^2}}=-jZ_2$$
(3.b)

$$\alpha_{z1}=\sqrt{\omega^2|\varepsilon_1|\mu_1+(\frac{m\pi}{a})^2+(\frac{n\pi}{b})^2}$$

$$\alpha_{z2}=\sqrt{\omega^2\varepsilon_2|\mu_2|+(\frac{m\pi}{a})^2+(\frac{n\pi}{b})^2}$$
(3.c)

As it is deduced from (3.b), the wave impedances of negative permittivity and permeability materials are inductive and capacitive, respectively. Thus, the wave propagation can be in principle obtained at any frequency by engineering the material parameters and adjusting the slab thicknesses correspondingly to satisfy the resonance condition. Because of the parametric dependence of eigenmode equation on monotonically increasing functions, only one resonance frequency is calculated for a certain phase shift per cell. This results only one frequency band to be obtained in the dispersion diagram for each transversal wave number in the case of lossless material parameters with low material dispersion. This property makes the novel cavity resonators with one resonance frequency for each transversal wave number to be designed by loading the negative permittivity material with the negative permeability material. However, this is not possible in the resonator designs by longitudinal pairing of two RHMs or one RHM with LHM [Engheta & Ziolkowski, 2006].

Another potential application is the design of monomode waveguides with arbitrary thick lateral dimension by loading the waveguide walls with the negative permeability and permittivity materials laterally instead of longitudinally [Alu & Engheta, 2003]. Another important issue is the design of a cavity resonator with negative material parameters at a predetermined resonance frequency, which can be derived from the eigenmode equation. Eigenmode equation indicates that the resonance frequency of a cavity resonator is only dependent on the ratio of each slab thickness rather than the total slab thickness unlike in the conventional resonators [Engheta, 2002]. The same result can also be concluded for the transmission line and resonator designs, which are based on pairing of LH and RH materials in the same unit cell but not of two RH materials. This property leads the compact subwavelength transmission media and resonators to be designed by pairing any slab thicknesses of negative permittivity and permability materials.This is also the main principle in the design of subwavelength guided wave structures with the lateral dimensions below the diffraction limits [Grbic & Eleftheriades,2004].

As a case study, one negative permittivity material of Drude type electric response and one negative permeability material of Lorentzian type magnetic response are periodically arranged inside a rectangular waveguide. The slab lengths of negative permittivity and permeability materials are 1mm and 2mm, respectively. The magnetic resonance, magnetic plasma frequency and loss parameter are 2.10 GHz, 2.21 GHz and 100 Hz for the negative permeability material. The electric plasma frequency and loss parameter are 10 GHz and 100 Hz for the negative permittivity material. As deduced from the material parameters, the material losses are taken into account in the model, however, kept quite small to have low transmission loss and observe the transmission band broadening. The dispersion diagram and Bloch impedance are analytically calculated and shown in Fig. 2 between the magnetic resonance and plasma frequencies of the negative permeability medium.

As it is deduced from Fig. 2a, the propagation constant is negative between 2.106 GHz and 2.146 GHz with 40 MHz bandwidth. It results approximately into Q factor of 53 in LH passband. However, in addition to this LH band, RH band is also obtained between 2.173 GHz and 2.21GHz with 37 MHz bandwidth. This RH passband has an approximate Q factor of 60 with the center frequency of 2.19GHz. There is 27 MHz bandgap between LH and RH bands, extending from 2.146 GHz to 2.173 GHz with the highest signal rejection level at the stop band frequency of 2.16 GHz. The emergence of these two bands is mainly related with the included material dispersion and loss. As shown in Fig. 2b, Bloch impedance is high ohmic at the lower edge and low ohmic at the higher edge with no reactive part. Thus, the composite material can be modeled as a combination of parallel and series resonant circuits with the resonance frequencies of 2.146 GHz and 2.173 GHz, respectively. In order to confirm this issue and the emergence of RH band at the higher frequencies, the equivalent circuit model of negative permittivity and permeability materials are illustrated in Fig. 3 [Engheta & Ziolkowski, 2006].

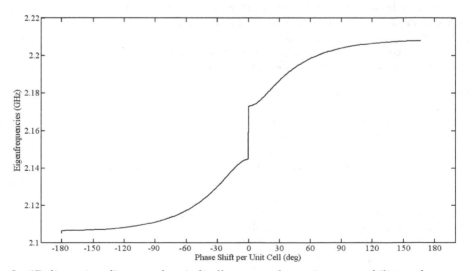

Fig. 2a. 1D dispersion diagram of periodically arranged negative permeability and permittivity materials

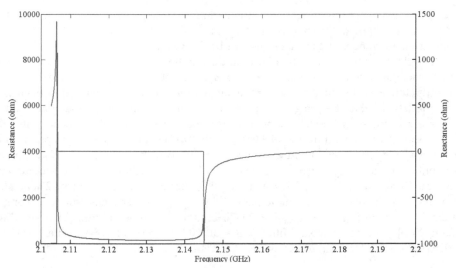

Fig. 2b. Resistance (red) and reactance (blue) of periodically arranged negative permeability and permittivity materials

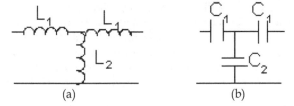

(a)   (b)

Fig. 3. Equivalent circuit model of one unit cell of (a) negative permittivity and (b) negative permeability materials

The periodic pairing of negative permittivity material with the negative permeability material results one unit cell to be modeled as

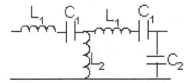

Fig. 4. Equivalent circuit model of one unit cell of periodically loaded negative permittivity and permeability materials

As deduced from Fig.4, for the frequencies larger than the series resonance frequency, the impedance of series branches is inductive. In addition, the resulting impedance of parallel branches are capacitive for the frequencies larger than the parallel resonance frequency. Thus, this circuit model has an equivalent form of series inductor loaded with the shunt capacitor as conventional RH transmission lines at the frequencies larger than the parallel and series resonance frequencies. This is the main reason why this composite material has an additional RH band at the frequencies larger than the LH band. In a similar manner, the equivalent circuit model of LH materials can be derived by the investigation of the same circuit model for the frequencies smaller than the parallel and series resonance frequencies. For this case, the circuit model has an equivalent form of series capacitor loaded with the shunt inductor. It is the dual of RH circuit model. One important conclusion from these circuit models is that the negative permittivity and permeability materials can be alternatively designed without relying on high lossy resonance phenomenon. In other words, rather than embedding the electrically small resonant metallic inclusions into the host material, the left handed feature can also be realized with low loss by periodic loading of conventional microstrip transmission lines with series capacitors and shunt inductors in planar microwave technology [Caloz & Itoh, 2005; Eleftheriades & Balmain, 2005]. Many microwave circuits have been implemented by using this strategy such as compact broadband couplers [Nguyen & Caloz 2007], broadband phase shifters [Eleftheriades & Balmain, 2005], compact wideband filters[Gil et al., 2007], compact resonant antennas [Lee et al., 2005, 2006; Schüßler et al., 2004; Sanada et al. 2004].

3. Spiral fractal resonator based compact band-stop filter

In this section, a compact, low insertion loss, high selective band-stop filter is explained. The filter is composed of two unit cells of electrically small artificial magnetic metamaterials, which have the topological form of fractal spiral resonators in [Palandoken & Henke, 2009]. The geometry of fractal spiral resonators is formed with the direct connection of two concentric Hilbert fractal curves of different dimensions as in a spiral form to excite the magnetic resonance. The operation principle of band-stop filter is based on the excitation of two electrically coupled fractal spiral resonators through direct connection with the feeding line.

The geometry of one unit cell of periodic artificial magnetic material is described shortly in Section 3.1 along with the reflection/transmission parameters and the resonant field pattern at the magnetic resonance frequency. In Section 3.2, the proposed filter topology is depicted with the geometrical and material parameters. The logical approach of the filter design is explained. In Section 3.3, the transmission and reflection parameters are illustrated to verify the desired filter performance and the operation principle is clarified in the light of simulated current distribution at the stopband frequency.

3.1 Structural design of fractal spiral resonator

The geometry of the artificial magnetic material is shown in Fig. 5. Each of the outer and inner rings are the mirrored image of the first order Hilbert fractal to form the ring shape. They are then connected at one end to obtain the spiral form from these two concentric Hilbert fractal curves. The marked inner section is the extension of the inner Hilbert curve so as to increase the resonant length due to the increased inductive and capacitive coupling between the different sections. The substrate material is standard 0.5 mm thick FR4 with dielectric constant 4.4 and tan(δ) 0.02. The metallization is copper. The copper line width and minimum distance between any two lines are 0.2 mm. The other geometrical parameters are L1= 2.2mm, L2= 0.8mm and L3= 1mm.

The unit cell size is ax = 5mm, ay =2mm, az = 5mm. Only one side of the substrate is structured with the prescribed fractal geometry while leaving the other side without any metal layer. Because two unit cells of this spiral fractal geometry is excited through the direct connection with the feeding line in the proposed band-stop filter, it is important to verify the magnetic resonance frequency as this frequency is the stopband frequency at which no field transmission is allowed.

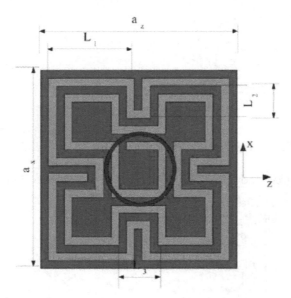

Fig. 5. Fractal spiral resonator geometry

In order to induce the magnetic resonance for the determination of stopband frequency, the structure has to be excited with out-off- plane directed magnetic field. Thus, in the numerical model, the structure is excited by z-direction propagating, x-direction polarized plane wave. Perfect Electric Conductor at two x planes and Perfect Magnetic Conductor at two y planes are assigned as the boundary conditions.

The resonance frequency and surface current distribution at the resonance frequency are numerically calculated with FEM based commercial software HFSS. The simulated S-parameters and surface current distribution are shown in Fig. 6 and Fig. 7, respectively.

Due to the spiralling form of surface current and the resulting out-off-plane directed magnetic field, the magnetic resonance is quite effective. Therefore, this electrically small structure can be regarded as a resonant magnetic dipole at 1.52 GHz. The transmission deep through one unit cell-thick artificial material in Fig.6 is effectively due to the depolarization effect of this magnetic dipole for the incoming field. This is the reason why this artificial magnetic material is regarded as a negative permeability material in a certain frequency band and the frequency of transmission deep is regarded as the magnetic resonance frequency of the electrically small structure. This frequency band is the desired band at which the proposed band-stop filter is designed to operate and the incoming field is exponentially attenuated in the propagation direction.

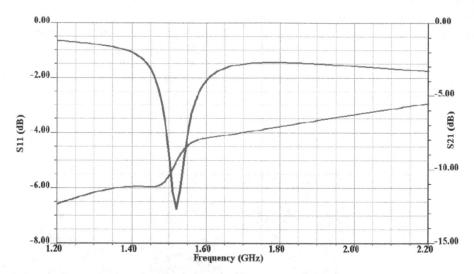

Fig. 6. Transmission (red) and reflection (blue) parameters of fractal spiral resonator

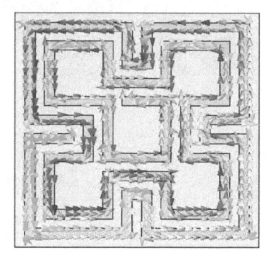

Fig. 7. Surface current distribution of fractal spiral resonator at 1.52 GHz

3.2 Band-stop filter design

The proposed band-stop filter is shown in Fig. 8. As it could be deduced from the filter topology, two fractal resonators are connected antisymmetrically along the x axis through the feeding line to have symmetrical return loss. In the filter model, the separation distance between x-direction oriented cells is 0.1 mm. Each unit cell is connected directly with the feeding line to increase the field coupling as shown with the red circles in Fig. 8 as a feeding method [Palandoken & Henke, 2010]. One conventional feeding method is to have resonator shaped slots in the ground plane for high field rejection [Marques et al., 2008]. However, this design method is not selected in order to suppress the resulting back radiation due to the slotted ground plane for lower insertion loss. In addition, this type of resonator feeding improves the filter selectivity in comparison to the proximity coupled feeding method. The separation distance between the unit cells and feeding line sections is 0.2 mm. Rather than using high lossy FR4, in the filter model low loss Rogers 4003 material with relative permittivity 3.38 and loss tangent 0.0027 is used. The main reason to use low-loss substrate instead of high-loss FR4 substrate is not only to decrease the insertion loss with low loss substrate but also with low permittivity material to increase the field coupling from the feeding line to the fractal spiral resonators.

The filter width (Wf) and length (Lf) are 10.1 mm and 5.4 mm, respectively. The width of each metallic line is 0.2 mm except the width of feeding line, which is 1.1 mm to excite both resonators effectively and couple each of the resonators with 50 Ω line impedance at both ports. The length of feeding line sections at each port is 5 mm. The total size of band-stop filter is 15.4 mm. The design principle of the proposed filter is to feed the fractal resonators directly through the feeding line with their resonant field distributions. This field distribution is the resonant field excited at the band edge of the bandgap in the Brilliouin diagram, which results into no transmission of the incoming wave because of its standing wave nature. As it could be noticed in Section 3.1, the resonance frequency of fractal spiral resonators is numerically calculated with the electrical coupling between the in-plane oriented cells and magnetic coupling between the out plane oriented cells in the transverse plane. The feeding method, which is currently exploited in this design, results these fractal spiral resonators to couple electrically as in the numerical model in Section 3.1. As a next step, the return and insertion loss of the band-stop filter is numerically calculated in addition to the resonant surface current distribution.

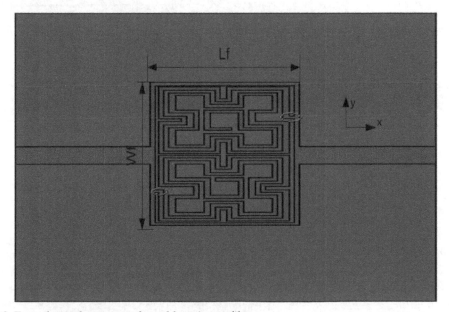

Fig. 8. Fractal spiral resonator based band-stop filter

3.3 Numerical results of fractal spiral resonator based band-stop filter

To validate the design concept, the performance of the band-stop filter is numerically calculated by using FEM based commercial software HFSS. The simulation results of the insertion and return losses are shown in Fig. 9. As shown in Fig. 9, the return loss is larger than 10 dB in two frequency bands of which is smaller than 0.75 GHz and which is larger than 1.87 GHz in the frequency span of 0.5-2.5 GHz. The insertion loss in the passband is better than 1 dB. The frequency rejection level is larger than 20 dB in the frequency band of 1.36-1.40 GHz and 27 dB at the center frequency, 1.38 GHz. The selectivity of the proposed band-stop filter is quite promising, which is 100 dB/GHz frequency selectivity with 3 dB reference insertion loss.

The physical size of the main filtering section is λo/40.18 x λo/21.48 at the center frequency, which is quite compact in comparison to the conventional stepped impedance or coupled line filters [Pozar, 2004]. On the other hand, there is no matching network, which is quite advantageous in the filter design to reduce the filter physical size. The total size of the filter even with the transmission line sections at the input and output ports is also compact, which is λo/10.85 x λo/14.46. The surface current distribution at the center frequency of the stopband, 1.38 GHz is shown in Fig. 10 to verify the design principle.

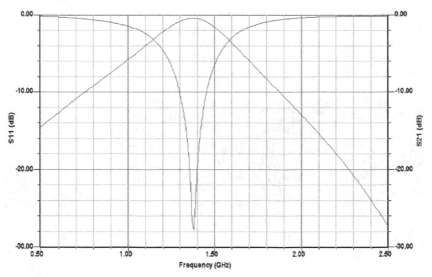

Fig. 9. Reflection(blue) and transmission (red) parameters of band-stop filter

As shown in Fig. 10, two fractal resonators are excited through the direct connection with the feeding line in their resonant current distribution. This type of resonator feeding leads these electrically small cells to have same direction directed magnetic dipole moments, which is actually the same field distribution for two electrically coupled spiral resonators. However, because of the additional electrical length due to the feeding line section for both of the spiral resonators, the center frequency of the stopband is lower than the resonance frequency of the fractal resonators.

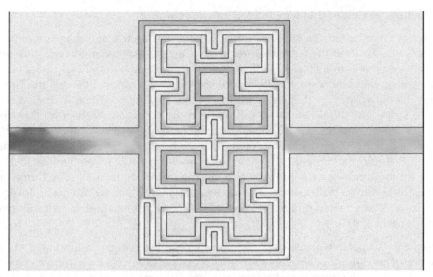

Fig. 10. Surface current distribution of fractal spiral resonator based BSF at 1.38GHz

The operation principle of the proposed filter is thus based on the suppression of the incoming field through the direct coupling of the feeding line to the spiral resonators at the resonant frequency. The resonant field, which is confined at the spiral resonators do not lead the magnetic dipole-like resonators to couple effectively to the electric dipole-like transmission line section at the second port. This reduces the transmission of the incoming field though the input feeding line and increases the reflection of the input signal at the first port.

As a result, in this section, the design of a compact, high selective, low insertion loss band-stop filter with two unit cells of novel magnetic metamaterial geometry is introduced. The physical dimensions of the designed band-stop filter are λo/10.85 x λo/14.46 with the additional transmission line sections at the center frequency of the stopband . No matching network is required in filter design, which is quite important to reduce the total filter size effectively. The insertion and return losses are numerically calculated and the current distribution at the center frequency of the stopband is illustrated. The proposed filter has satisfactory insertion loss, which is better than 1 dB. The selectivity of the filter is 100 dB/GHz, which is quite suitable to be used as band-stop filter in modern communication systems.

4. Thin wire loaded spiral resonator based compact band-pass filter

In this section, the design of a compact, LHM-based band-pass filter (BPF) is explained. The band-pass filter is composed of two unit cells of electrically small LHMs, which have the same geometry of thin wire loaded spiral resonators in [palandoken et al., 2009]. The LHM geometry is based on the direct connection of thin wire with the spiral resonator in order to enhance the magnetic field coupling inbetween in addition to the increased electrical length for reduced resonance frequency. The operation principle of BPF is based on the excitation of two coupled LHM resonators at two eigenfrequencies through direct connection with the feeding line.

4.1 Structural design of thin wire loaded spiral resonator

LHM behavior relies on the simultaneous excitation of electric and magnetic dipole-like electrically small cells at the resonance frequency. One well-known miniaturization method of such resonators is to increase the field coupling among the individual resonators. This design strategy is chosen for the LHM cell exploited in current BPF design. The LHM geometry is shown in Fig. 11 along with the boundary conditions and excitation sources. The wire strips and spiral resonators (SR) are directly connected with each other on both sides of the substrate. Further, instead of SRRs as in the original proposal of artificial magnetic material, SRs are used, which have half the resonance frequency of SRRs [Baena et al., 2004, 2005]. In the design, the geometrical parameters of the front and back side unit cells are the same, except a 0.6 mm shorter wire strip length on the front side. Different strip wire lengths lead to a smaller resonance frequency and larger bandwidth [palandoken et al., 2009]. The substrate material is nonmagnetic FR4-Epoxy with a relative permittivity of 4.4 and loss tangent of 0.02.

The frequency dispersive property of FR4 has not been taken into account in the numerical calculation. The validity of the model is illustrated by retrieving the effective constitutive parameters from S parameters and by the opposite direction of group and phase velocity in [palandoken et al., 2009].

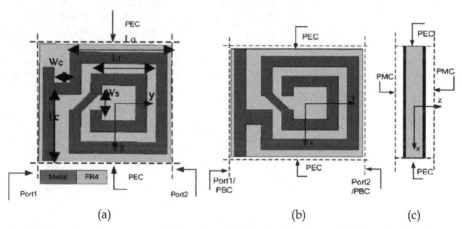

(a)   (b)   (c)

Fig. 11. LHM unit cell geometry. (a) Front and (b) back side of one LHM unit cell with indicated geometrical parameters in [palandoken et al., 2009].

4.2 Band-pass filter design

The band-pass filter is shown in Fig. 12. As it could be implied from the filter topology, two LHM resonators are connected antisymmetrically along the y axis through the feeding line in order to have symmetrical reflection parameter. In the filter model, the separation distance between x-direction oriented cells is 0.2 mm. Each unit cell is connected directly with the feeding line to excite each of LHM resonators effectively and couple the excited LHM resonator electrically and magnetically with the another resonator depending on the mode of excitation in addition to the extended feeding line sections as an effective feeding method. LHM resonator shaped slots in the ground plane are not exploited in the current band-pass filter due to the resulting back radiation from the slotted ground plane. The separation distance between one unit cell and feeding line sections in y-direction is 0.8 mm. One LHM resonator is shifted from the another resonator along y-direction with the distance of 0.4 mm to obtain optimum field intercoupling. Instead of using high lossy FR4, in the filter model low loss Rogers 5880 material with the relative permittivity 2.2 and loss tangent 0.0009 is used. The main reason to use low-loss substrate instead of high-loss FR4 substrate is not only to decrease the insertion loss with low loss substrate but also with low permittivity material to increase the field coupling from the first resonator into the second resonator in addition to the field coupling from the feeding line to each of LHM resonators.

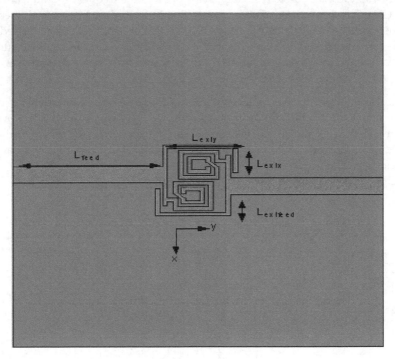

Fig. 12. Thin wire loaded spiral resonator based BPF

The lengths of extended feeding line in x- (L_{extx}) and y-directions (L_{exty}) are 2.5mm and 4.9mm, respectively. The width and length (L_{feed}) of feeding line are 1.5mm and 9.8mm, respectively. The direct connection length of the feeding line, ($L_{extfeed}$) in the extended sections is 1.9mm. The width and length of Rogers substrate are 30 mm and 24.2mm, respectively. The total size of band-pass filter is 6.4mm x 5.4 mm.

The design principle of the proposed filter is to feed one of the LHM resonators directly through the feeding line with its resonant field distribution. The highly concentrated field distribution results the directly excited first resonator to couple to the second LHM resonator electrically and magnetically depending on the excited 0 and 180 modes. The excited field in the first LHM resonator is enhanced by the magnetic field of the extended section of the feeding line. The feeding method, which is currently exploited in this design, results these LHM resonators to couple feeding line in an optimal manner. As a next step, the return and insertion loss of the band-pass filter is numerically calculated in addition to the resonant surface current distribution at two eigenfrequencies.

4.3 Numerical results of thin wire loaded spiral resonator based band-pass filter

To validate the filter design concept, the reflection and transmission parameters of band-pass filter are numerically calculated by using FEM based commercial software HFSS. The numerical results are shown in Fig. 13. As deduced from Fig. 13, the return loss is larger than 10 dB with the insertion loss smaller than 1.3 dB in the frequency band from 4.49GHz upto 4.92 GHz. The lowest insertion loss is 0.4 dB at 4.83GHz. The filter selectivities , roll-off factors, are approximately 400 dB/GHz and 62 dB/GHz at the lower and higher edge of the passband, respectively. The filter response is quite similar to the transfer function of Chebychev filter due to existing passband ripples and steep roll-off from the passband edges to the stop-band frequencies.

The physical size of the filtering section without the feeding line sections is $\lambda o/10 \times \lambda o/12$ at the center frequency of 4.7GHz. It is quite compact in comparison to the conventional stepped impedance or coupled line filter designs. One important advantage of the current filter design is to have no additional matching network, which reduces the filter physical size significantly. The surface current distributions at two resonance frequencies, 4.52 GHz and 4.83 GHz, are shown in Fig. 14 to verify the design principle.

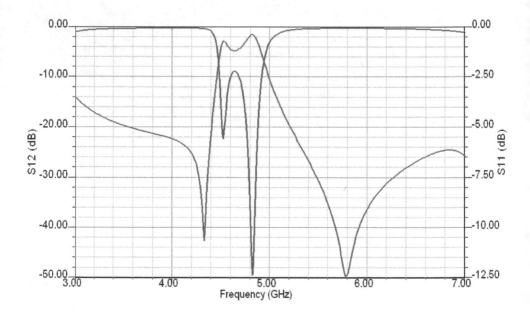

Fig. 13. Reflection(blue) and transmission (red) parameters of band-pass filter

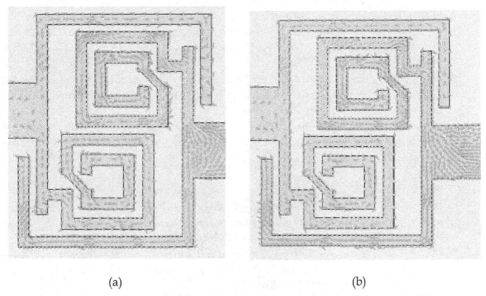

(a)   (b)

Fig. 14. Surface current distribution of thin wire loaded spiral resonator based BPF at
(a) 4.52 GHz and (b) 4.83GHz

As shown in Fig. 14, one of LHM resonators is directly fed through the feeding line
without any metallic connection from the excited resonator to the another LHM resonator.
The resonant field distribution of LHM resonator and extended feeding line leads the
incoming field to be coupled from the input port to the output port in 0 mode at 4.52GHz
and 180 (π) mode at 4.83 GHz. The electric coupling at the lower resonance frequency
results both LHM resonators and extended feeding line sections to radiate in comparison
to the magnetic coupling at the higher resonance frequency. This is the main reason why
the insertion loss is higher at the lower resonance frequency than that at the higher
resonance frequency. The transmission principle of incoming field in the passband is
based on the excitation of output LHM resonator by the electric and magnetic coupling of
the resonant field excited in the directly fed resonator at the lower and higher resonance
frequency, respectively.

As a result, in the last section of this chapter, the design of a compact band-pass filter with
two unit cells of LHM geometry is explained. The physical dimensions of band-pass filter
are λo/10 x λo/12 at the center frequency of 4.7GHz. There is no need of additional
matching network in filter design, which shrinks the total filter size effectively. The insertion
and return losses are numerically calculated. The current distributions at two resonance
frequencies are illustrated. The proposed filter has satisfactory insertion loss, which is
smaller than 1.2 dB in the frequency band between 4.49GHz and 4.92 GHz with the
bandwidth of 430MHz. The filter selectivities calculated with reference to 3dB insertion loss
value are 400 dB/GHz and 62 dB/GHz at the lower and higher edge of the passband,
respectively.

5. References

N. G. Alexopoulos, C. A. Kyriazidou, and H. F. Contopanagos, "Effective parameters for metamorphic materials and metamaterials through a resonant inverse scattering approach," *IEEE Trans. Microw. Theory Tech.*, vol. 55, no. 2, pp. 254–267, Feb. 2007

A. Alù and N. Engheta, "Pairing an epsilon-negative slab with a mu-negative slab: Resonance, tunneling and transparency," IEEE Trans. Antennas Propag., vol. 51, no. 10, pp. 2558–2571, Oct. 2003.

A. Alù and N. Engheta, "Guided modes in a waveguide filled with a pair of single-negative(SNG), double-negative (DNG), and/or double-positive (DPS) layers," IEEE Trans. Microw. Theory Tech., vol. 52, no. 1, pp. 199–210, Jan. 2004.

J. Baena, R. Marqués & F. Medina, "Artificial magnetic metamaterial design by using spiral resonators," Phy. Rev. B, vol. 69, pp. 0144021–0144025, 2004

C.Caloz and T. Itoh, *Electromagnetic Metamaterials: Transmission Line Theory and Microwave Applications*, Piscataway, NJ: Wiley- IEEE, 2005

X. Chen, T. M. Grzegorczyk, B.-I.Wu, J. Pacheco, Jr., and J. A. Kong, "Robust method to retrieve the constitutive effective parameters of metamaterials," *Phys. Rev. Lett. E*, vol. 70, pp. 0166081–0166087, 2004.

G. V. Eleftheriades and K. G. Balmain, *Negative Refraction Metamaterials: Fundamental Principles and Applications*, New York:Wiley Interscience, 2005

N. Engheta, "An idea for thin subwavelength cavity resonators using metamaterials with negative permittivity and permeability," IEEE Antennas Wireless Propag. Lett., vol. 1, no. 1, pp. 10–13, 2002

N. Engheta and R. W. Ziolkowski, *Metamaterials Physics and Engineering Explorations*, Eds. New York: Wiley/IEEE, 2006.

M. Gil, J. Bonache, J. Garcia-Garcia, J. Martel & F. Martin, "Composite right/left-handed metamaterial transmission lines based on complementary split-rings resonators and their applications to very wideband and compact filter design," *IEEE Trans. Microw. Theory Tech.*, vol. 55, no. 6, pp. 1296–1304, Jun. 2007.

A. Grbic and G. V. Eleftheriades, "Overcoming the diffraction limit with a planar left- handed transmission lines," Phys. Rev. Lett., vol. 92, no. 11, p. 117 403, Mar. 2004.

A. Grbic and G. V. Eleftheriades, "Experimental verification of backward-wave radiation from a negative refractive index metamaterial," J.Appl. Phys., vol. 92, pp. 5930–5935, Nov. 2002.

P. Kolinko and D. R. Smith, "Numerical study of electromagnetic waves interacting with negative index materials," Opt. Express, vol. 11, pp. 640–648, Apr. 2003

J. A. Kong, B.-I. Wu, and Y. Zhang, "A unique lateral displacement of a Gaussian beam transmitted through a slab with negative permittivity and permeability," Microwave Opt. Technol. Lett., vol. 33, pp. 136–139, Mar. 2002.

C.-J. Lee, K. M. K. H. Leong, and T. Itoh, "Composite right/left-handed transmission line based compact resonant antennas for RF module integration," *IEEE Trans. Antennas Propag.*, vol. 54, no. 8, pp. 2283–2291, Aug. 2006.

C.-J. Lee, K. M. K. H. Leong, and T. Itoh, "Design of resonant small antenna using composite right/left handed transmission line," in *Antennas Propag. Soc. Int. Symp.*, 2005, vol. 2B

Ricardo Marqués, Ferran Martín, Mario Sorolla, *Metamaterials with negative parameter:theory, design, and microwave applications*, John Wiley & Sons, 2008

H. V. Nguyen and C. Caloz, "Generalized coupled-mode approach of metamaterial coupled line couplers: Coupling theory, phenomenological explanation and experimental demonstration," *IEEE Trans. Microw.Theory Tech.*, vol. 55, no. 5, pp. 1029–1039, May 2007

M. Palandöken, A. Grede & H. Henke, "Broadband microstrip antenna with Left-handedMetamaterials, "IEEE Trans. Antennas Propag., vol. 57, no.2, pp. 331–338, Feb. 2009

M. Palandöken, H. Henke, "Fractal spiral resonator as magnetic metamaterial, "Applied Electromagnetics Conference, pp. 1–4, 2009

M. Palandöken, H. Henke, "Compact LHM-based band-stop filter", Mediterranean Microwave Symposium, pp. 229–231, Aug. 2010

J. B. Pendry, A. J. Holden, D. J. Robbins & W. J. Stewart, "Magnetism from conductors and enhanced nonlinear phenomena," IEEE Trans. Microw. Theory Tech., vol. 47, no. 11, pp. 2075–2081, Nov. 1999

J. B. Pendry, A. J. Holden, D. J. Robbins & W. J. Stewart, "Low-frequency plasmons in thin wire structures," J. Phys., Condens.Matter, vol. 10, pp. 4785–4809, 1998

David Pozar, *Microwave Engineering* , 2004, Wiley

A. Sanada, M. Kimura, I. Awai, C. Caloz, and T. Itoh, "A planar zeroth- order resonator antenna using a left handed transmission line," in *34th Eur. Microw. Conf.*, Amsterdam, The Netherlands, 2004, pp. 1341–1344.

M. Schüßler, J. Freese, and R. Jakoby, "Design of compact planar antennas using LH-transmission lines," in *Proc. IEEE MTT-S Int. Microw. Symp.*, 2004, vol. 1, pp. 209–212.

D. R. Smith and N. Kroll, "Negative refractive index in left-handed materials," Phys. Rev. Lett., vol. 85, pp. 2933–2936, Oct. 2000.

D. R. Smith, D. C. Vier, N. Kroll, and S. Schultz, "Direct calculation of permeability and permittivity for a left- handed metamaterial," *App.Phys. Lett.*, vol. 77, no. 14, pp. 2246–2248, Oct. 2000

D. R. Smith, D. C. Vier, T. Koschny, and C. M. Soukoulis, "Electromagnetic parameter retrieval from inhomogeneous metamaterials," *Phys. Rev. E*, vol. 71, pp. 0366171–03661711, 2005

R.W. Ziolkowski and A. Kipple, "Application of double negative metamaterials to increase the power radiated by electrically small antennas," IEEE Trans. Antennas Propag., vol. 51, no. 10, pp. 2626–2640, Oct. 2003

R. W. Ziolkowski, "Pulsed and CW Gaussian beam interactions with double negative metamaterial slabs," Opt. Express, vol. 11, pp. 662–681, Apr. 2003

V. G. Veselago, "The electrodynamics of substances with simultaneously negative values of ε and μ," Sov. Phys.—Usp., vol. 47, pp. 509–514, Jan.–Feb. 1968.

7

Magnetically Tunable Unidirectional Electromagnetic Devices Based on Magnetic Surface Plasmon

Shiyang Liu[1], Huajin Chen[1], Zhifang Lin[2] and S. T. Chui[3]
[1]Institute of Information Optics, Zhejiang Normal University
[2]State Key Laboratory of Surface Physics (SKLSP) and Department of Physics,
Fudan University
[3]Department of Physics and Astronomy, University of Delaware
[1,2]China
[3]USA

1. Introduction

Plasmonic metamaterials are composite, artificial materials, consisting of metallic resonant building blocks designed with state-of-the-art configurations. Many exotic phenomena not occurring in nature such as negative refraction (Burgos et al., 2010; Dolling et al., 2006; Lezec et al., 2007; Liu et al., 2008; Pendry, 2000; Shalaev, 2007; Shelby et al., 2001; Smith et al., 2000; Valentine et al., 2008; Veselago, 1968; Zhang et al., 2005), subwavelength imaging (Fang et al., 2005; Liu et al., 2007; Pendry, 2000; Taubner et al., 2006), cloaking (Ergin et al., 2010; Leonhardt, 2006; Lai et al., 2009; Li et al., 2008; Liu et al., 2009; Pendry et al, 2006; Schurig et al., 2006), and so on can be observed in the systems made of such materials. A particular characteristic of the system is the excitation of surface plasmon polaritons (SPPs) (Barnes et al., 2003; Giannini et al., 2010; Noginov et al., 2008; Zayats et al., 2005), which originates from the coupling of the electromagnetic (EM) wave to the free electron oscillation in metallic surface. The EM waveguiding mediated by the SPPs has gained more and more attention from theorists, experimentalists, and engineers due to its promising applications in integrated optical circuit, optical storage, biosensing, and even medical therapy (Engheta, 2007; Ozbay, 2006; Zia et al., 2006). A series of plasmonic components have been proposed and realized, such as plasmonic waveguides, beam splitter, sharp bends and so on. Very recently, based on the excitation of the SPPs the plasmonic Luneburg and Eaton lens are even realized experimentally (Zentgraf, 2011).

So far, plasmonics relevant phenomena have been extensively investigated from both the theoretical and the experimental perspectives. The symmetry of the Maxwell's equations with respect to its electric and magnetic part suggests that the magnetic analogue of the surface plasmon, "magnetic surface plasmon" (MSP) (Gollub et al., 2005; Liu et al., 2008), can also be excited. It originates from the coupling of EM wave to the collective resonance of spin wave, and thus can be observed in the magnetic system. However, the relevant issues are not examined elaborately as the case for the surface plasmon in metallic materials. In addition, there appeared burgeoning activities in exploring the phenomena resulting from the time reversal symmetry (TRS) breaking for photons. Among others, the EM one-way edge modes analogous to quantum Hall edge states were discussed by Haldane and Raghu (Haldane & Raghu, 2008), and later by Wang and co-workers (Wang et al., 2008; 2009). Recently, the self-guiding unidirectional EM edge states are also realized theoretically (Ao et al., 2009) and experimentally (Poo et al., 2011). In the metamaterials designed with ferrite materials, the magnetic response is intrinsic, accordingly, termed as magnetic metamaterials (MMs), where the MSP resonance can be excited. Besides, the TRS breaking relevant phenomenon for the photons can also be observed (Liu et al., 2010; 2011; Wang et al., 2008). In particular, we can find that the MMs can mold the reflection in a dramatic manner (Liu et al., 2011).

Similar to the electric surface plasmon (ESP) resonance occurring when its permittivity $\varepsilon = -1$ in metallic rod. For a ferrite rod, the MSP resonance occurs when its effective permeability $\mu = -1$ (Liu et al., 2008). Due to the TRS breaking nature in ferrite materials, there exist unequal amounts of states with opposite angular momenta for the scattered field. Near the MSP resonance, the angular momenta contents are dominated by one sign, with that for the other sign almost completely suppressed. Accordingly, a giant circulation (clockwise or anticlockwise) of energy flow develops, so only the energy flow in one direction is supported (Chui & Lin, 2007). Consequently, in this work we consider the phenomenon resulting from the combined action of the MSP resonance and the TRS breaking (Liu et al., 2011). In addition, the working frequency can be controlled by an external magnetic field (EMF), facilitating the design of the practical EM devices.

The present chapter is organized as follows. In the second part, we give a brief introduction on the Mie theory on the ferrite rod and the multiple scattering theory used in the numerical simulations. Then, we present the reflection behavior due to the MSP resonance and its physical origin. The dependence of the behavior on the working frequency and the source-interface separation is also examined. Following this part, we show a design of a one-way EM waveguide (OEMW) based on this effect. Most importantly, the robustness of the OEMW against defect, disorder, and inhomogeneity of the EMF are examined. The manipulability of the working frequency is demonstrated as well by tuning the EMF. Some other complicate EM devices such as a sharp beam bender and a beam splitter are also designed. In addition, all the designed EM devices are shown to be operable even in the deep subwavelength scale. Our results are summarized in the conclusion part.

2. Theoretical approach

To design the MM, we use single crystal yttrium-iron-garnet (YIG) ferrite rods as building blocks. In our case, the ferrite rods are arranged periodically as a square lattice in the air with lattice constant a and the radius of the ferrite rod r_s. The rod axis are oriented along the z direction, corresponding to the direction of the EMF. When fully magnetized, the magnetic permeability can be written in the form (Pozar, 2004; Slichter, 1978)

$$\widehat{\mu} = \begin{pmatrix} \mu_r & -i\mu_\kappa & 0 \\ i\mu_\kappa & \mu_r & 0 \\ 0 & 0 & 1 \end{pmatrix}, \qquad \widehat{\mu}^{-1} = \begin{pmatrix} \mu_r' & -i\mu_\kappa' & 0 \\ i\mu_\kappa' & \mu_r' & 0 \\ 0 & 0 & 1 \end{pmatrix}, \tag{1}$$

with

$$\mu_r = 1 + \frac{\omega_m(\omega_0 - i\alpha\omega)}{(\omega_0 - i\alpha\omega)^2 - \omega^2}, \quad \mu_\kappa = \frac{\omega_m\omega}{(\omega_0 - i\alpha\omega)^2 - \omega^2}, \quad \mu_r' = \frac{\mu_r}{\mu_r^2 - \mu_\kappa^2}, \quad \mu_r' = \frac{-\mu_\kappa}{\mu_r^2 - \mu_\kappa^2},$$

where $\omega_0 = \gamma H_0$ is the resonance frequency with $\gamma = 2.8$ MHz/Oe the gyromagnetic ratio; H_0 the sum of the EMF applied in z direction and the shape anisotropy field (Pozar, 2004), $\omega_m = 4\pi\gamma M_s$ is the characteristic frequency with $4\pi M_s = 1750$ Oe the saturation magnetization, and α is the damping coefficient of the ferrite. In the calculation of photonic band diagram, we set $\alpha = 0$ (Wang et al., 2008), and in the simulation $\alpha = 3 \times 10^{-4}$. The relative permittivity of the ferrite rods is taken as $\varepsilon = 15 + 3 \times 10^{-3}i$. In our work, the transverse magnetic (TM) mode is considered.

To make our results reliable and convergent efficiently, we have used the Mie theory to solve the scattering properties of a single ferrite rod. For an incident EM wave with TM polarization, it can be expanded into vector cylindrical wave functions (VCWFs) (Bohren & Huffman, 1983)

$$E_0(j) = \sum_n E_n q_n^{(j,j)} N_n^{(1)}(k_b, r_j), \tag{2}$$

where j suggests that the $j-$th coordinate is used, indicating that the EM fields are expanded around the $j-$th ferrite rod, $N_n^{(1)}(k_b, r_j)$ is the VCWF with k_b the wavenumber in the background medium, satisfying $k_b^2 = \omega^2 \varepsilon_b \mu_b$, ε_b and μ_b correspond, respectively, to the permittivity and permeability of the background medium, $E_n = i^n |E_0|$ with $|E_0|$ the amplitude of the incident EM wave. For plane wave, the expansion coefficients can be easily obtained

$$q_n^{(j,j)} = e^{-in\phi_\kappa} \exp(ik_b \cdot d_{0j}), \tag{3}$$

where $k_b = \kappa \cos\phi_\kappa e_x + \kappa \sin\phi_\kappa e_y$ is the wave vector of the incident EM wave with $\kappa = |k_b|$, $d_{0j} = r - r_j$, corresponding to the position of the $j-$th ferrite rod in the $0-$th coordinate. The scattered EM wave can be expanded into VCWFs as well in the form

$$E_s(j) = -\sum_n E_n b_n^{(j,j)} N_n^{(3)}(k_b, r_j), \tag{4}$$

In Eqs. (2) and (4), the VCWFs are defined according to

$$N_n^{(J)}(k, r) = z_n^{(J)}(kr)e^{in\phi}e_z, \tag{5}$$

where

$$z_n^{(J)}(kr) = \begin{cases} J_n(kr) & \text{for } J = 1, \\ H_n^{(1)}(kr) & \text{for } J = 3, \end{cases} \tag{6}$$

In Eq. (6), $J_n(kr)$ and $H_n^{(1)}(kr)$ correspond, respectively, to the first kind of Bessel function and the first kind of Hankel function.

The EM wave inside the ferrite rod can also be expanded into VCWFs, but the form is so complicated. Therefore, to save the space we do not give the corresponding result. By matching the boundary conditions, we can solve the problem and obtain the Mie coefficients, which can connect the unknown scattering coefficients with the given expansion coefficients of the incident wave in the form

$$b_n^{(j,j)} = S_n^{(j)} q_n^{(j,j)}, \tag{7}$$

where $S_n^{(j)}$ is the Mie coefficients of the $j-$th ferrite rod (Chen et al., 2007; Chui & Lin, 2007; Eggimann, 1960)

$$S_n = \frac{\frac{\mu_s}{\mu_b} J_n'(x) - J_n(x) \left[\frac{m_s^2}{m_s'} D_n(m_s'x) + \frac{n\mu_k'}{x} \right]}{\frac{\mu_s}{\mu_b} H_n^{(1)'}(x) - H_n^{(1)}(x) \left[\frac{m_s^2}{m_s'} D_n(m_s'x) + \frac{n\mu_k'}{x} \right]}. \tag{8}$$

In Eq. (8), the superscripts $"/"$ of $J_n(x)$ and $H_n(x)$ denote the derivatives with respect to the argument $x = k_b r_s$, the other parameters $k_s^2 = \omega^2 \epsilon_s \mu_s$, $m_s = k_s/k_b$, $m_s' = m_s/\sqrt{\mu_r'}$, and $D_n(m_s'x) = J_n'(m_s'x)/J_n(m_s'x)$.

By combining the Mie theory with the multiple scattering theory (Felbacq et al., 1994; Leung & Qiu, 1993; Liu & Lin, 2006; Wang et al., 1993), we can handle the scattering problem for a system consisting of multiple ferrite rods. Then, the scattering coefficients around each ferrite rod can be obtained by solving the linear equations

$$b_n^{(j)} = S_n^{(j)} \left[q_n^{(j,j)} - \sum_{l \neq j} \sum_m A_{nm} b_m^{(l)} \right]. \tag{9}$$

In Eq. (9), A_{nm} is the translational coefficient (Chew, 1995)

$$A_{nm} = i^{n-m} H_{n-m}^{(1)}(kd_{lj}) \exp\left[-i(n-m)\phi_{lj} \right], \tag{10}$$

where $d_{lj} = r_l - r_j$ and (d_{lj}, ϕ_{lj}) is the polar coordinate of the position vector d_{lj}. By setting $q_n^{(j,j)} = 0$ and find the corresponding eigenmodes, we can also calculate the photonic band diagrams.

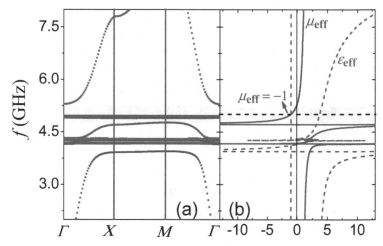

Fig. 1. (a) The photonic band diagram for an MM of square lattice with lattice constant $a = 8$ mm and under an EMF of $H_0 = 900$ Oe. (b) The retrieved effective constitutive parameters where the blue dashed line and red solid line correspond to the effective permittivity ε_{eff} and the effective magnetic permeability μ_{eff}, respectively.

3. MSP resonance in MM

The MM considered are made of single crystal YIG rods arranged periodically in a square lattice with the lattice constant $a = 8$ mm and the rod radius $r = \frac{1}{4}a = 2$ mm. Firstly, let's examine the origin of the MSP resonance. It depends on the resonance of the each single ferrite rod. The coupling of the neighboring MSP resonant states of each single ferrite rod results in the resonance of the whole system. For a ferrite rod with the permeability given by Eq. (1), the MSP resonance occurs at a frequency

$$f_s = \frac{1}{2\pi} \left(\omega_0 + \frac{1}{2}\omega_m \right), \tag{11}$$

corresponding to the case when $\mu_r + \mu_\kappa = -1$ (Liu et al., 2008), which can be regarded as the effective permeability of a single ferrite rod. In the present case, the applied EMF field is such that $H_0 = 900$ Oe, yielding a MSP resonance at $f_s = 4.97$ GHz. To gain a better understanding of the MSP resonance, we present the photonic band diagram in Fig. 1 (a). At the working frequency $f = 5$ GHz (corresponding the wavelength 60 mm) the wavelength is nearly 8 times of the lattice constant a so that the MM can be considered as an effective medium. With the effective-medium theory developed in our previous work (Jin et al., 2009), we can retrieve the effective permeability tensor,

$$\mu_e = \begin{pmatrix} \mu & -i\mu' & 0 \\ i\mu' & \mu & 0 \\ 0 & 0 & 1 \end{pmatrix}. \tag{12}$$

Then, the effective permeability μ_{eff} for TM mode is given by $(\mu^2 - \mu'^2)/\mu$, which is shown in Fig. 1(b). By comparing the photonic band diagram and the effective constitutive parameters μ_{eff} and ε_{eff}, we can find that double positive effective parameters correspond to the photonic bands, single negative effective parameter corresponds to the photonic band gap (PBG), while the resonances correspond to the flat bands. Near the frequency corresponding to the flat bands in Fig. 1(b) (at the MSP resonance frequency $f_s = 4.97$ GHz) the effective magnetic permeability is nearly equal to -1 as is marked by the black dashed line, consistent with the analysis on the single ferrite rod. For this reason, the MSP resonance can be considered as the magnetic analogue of the surface plasmon resonance inhabited in the metallic materials. In addition, around the MSP resonance two PBGs come into existence as denoted by the yellow stripes in Fig. 1(a) where $\mu_{\text{eff}} < 0$, in the vicinity of the MSP resonance. This is also the working frequency range we select to examine the corresponding reflection behavior.

4. Molding reflection with MM slab

To explore the physical consequence resulting from the combined action of the MSP resonance and the TRS breaking, we have examined the reflection behavior of a TM wave excited by a line source working at a frequency near f_s and located near an MM slab. Typical results are demonstrated in Fig. 2, where the line source oscillating at $f = 5$ GHz is located a distance $a = 8$ mm away from a four-layer MM slab. The total electric field, the scattered electric field, together with the x-component of the Poynting vector are plotted in panels (a), (b), and (c), respectively, where a sharply asymmetric reflection (SAR) can be observed. On the left hand side (LHS) of the line source, the scattered field substantially cancels the incoming field, resulting in a darkened region near the MM surface. On the right hand side (RHS) of the line source, the scattered field significantly enhances the EM field, giving rise to a brightened region near the MM surface. Since the working frequency is selected in the PBG as can be seen from the photonic band diagram in Fig. 1 (a), the EM wave can not propagate inside the MM slab. The vanishment of the total field inside the MM slab indicates that the incident and scattered fields inside the MM have a π phase difference. For the scattered wave, on the LHS of the line source, the bright fringes both inside and outside the MM [shown in Fig. 2 (b)] remains at the same positions. This indicates that the scattered waves on the LHS are continuous in phase near the surface of MM. As the scattered field cancels the incident field

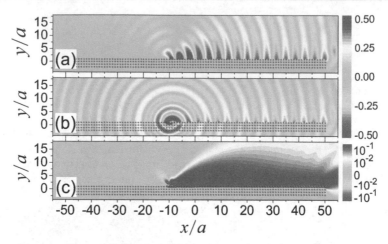

Fig. 2. The patterns of the total electric field (a), the scattered electric field (b), and the x-component of the Poynting vector (c) for a line source operating at a frequency $f = 5$ GHz close to the MSP resonance. The lattice constant of the MM slab is $a = 8$ mm and the line source is located $1.0a$ away from the surface of the MM slab.

inside the MM slab, it therefore also attenuates the incident field outside the MM slab. The situation is quite different on the RHS of the line source. As can be seen from Fig. 2 (b), the bright fringe outside the MM slab is at the same position as the dark fringe inside the MM slab near the interface. There appears a nearly half-wavelength mismatch for scattered waves inside and outside the MM slab. This suggests a phase change around π occurs inside and outside the MM slab. As a consequence, while the scattered field cancels the incident field inside the slab due to the phase mismatch, it considerably enhances the EM field near surface outside the MM slab owing to in-phase interference. In this manner, we can understand the SAR effect phenomenologically.

As can be seen from Fig. 1, the working frequency is selected near the MSP resonance, which plays a crucial role for the phenomenon occurring in the system. The excitation of the MSP resonance can lead to the appearance of the unidirectional circulation of the energy flow as schematically shown in Fig. 3. Accordingly, for a line source located near the surface of the MM slab, the leftward energy flow is inhibited while the rightward energy flow is supported

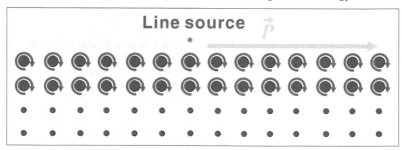

Fig. 3. The simple physical picture illustrating the excitation of the one-way circulating MSP band states, which is responsible for the occurrence of the SAR effect shown in Fig. 2.

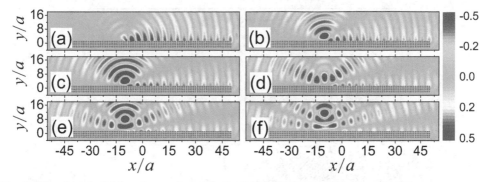

Fig. 4. The electric field patterns showing the SAR effect corresponding to different source-interface separations d. Panels (a), (b), (c), (d), (e), and (f) correspond to the cases when $d = 1a, 2a, 3a, 5a, 7a$, and $10a$, respectively. All the other parameters are the same as those in Fig. 1. The working frequency is selected at $f = 5\,\text{GHz}$. The MM slab is the same as that in Fig. 2.

and reinforced. For this reason, we can observe the SAR effect shown in Fig. 2. Interestingly, by reversing the orientation of the EMF the circulation of the energy flow can also be reversed. Therefore, the EM property relevant to the SAR effect is magnetically tunable.

4.1 Separation dependence of the SAR effect

As aforementioned, the SAR effect originates from the MSP resonance. Therefore, it is necessary to examine the dependence of the SAR effect on the source-interface separation d. The corresponding results are shown in Fig. 4 where panels (a), (b), (c), (d), (e), and (f) correspond to the electric field patterns when $d = 1a, 2a, 3a, 5a, 7a$, and $10a$, respectively. The operating frequency is $f = 5\,\text{GHz}$, all the other parameters of the MM slab are also the same as those in Fig. 2. When the line source is close to the interface of MM slab a dramatic SAR effect can be observed. With the increase of the separation d, the SAR effect becomes weaker and weaker, namely, more and more electric field is scattered into the outside space. When the line source is far enough from the interface ($d \approx \lambda$), the electric field near the interface becomes very weak and the electric field pattern is nearly symmetric as can be observed in Figs. 4 (e) and (f). This result can be understood from the following approximate physical picture. The incoming electric field at a position r is given by

$$E_i(r) = \exp[ik \cdot (r - e_y d)]/|r - e_y d|. \tag{13}$$

The scattered electric field is approximately of the spatial dependence of that from an image term

$$E_s(r) = s\exp[ik \cdot (r + e_y d)]/|r + e_y d|, \tag{14}$$

where $s = \pm 1$, depending on whether one is on the LHS or RHS of the source. For small d, there is an obvious cancellation between these two terms. As d is increased, this cancellation only remain at $y = 0$. But for other values of y, this cancellation becomes too much weaker.

4.2 Frequency dependence of the SAR effect

The MSP resonance occurs at a specified frequency. Accordingly, the SAR effect should be also dependent on the selection of the working frequency. In Fig. 5 (a), we present the photonic

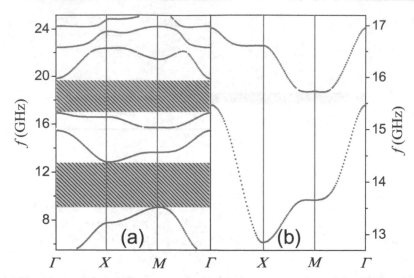

Fig. 5. (a) The photonic band diagram at the higher frequency where three PBGs arising from the bragg scattering can be observed. (b) The amplified view of the photonic band diagram around the second PBG. All the parameters involved are the same as those in Fig. 1.

band diagrams at higher frequencies, where three PBGs are identified. We first examine the reflection behavior at the frequency $f = 9.5$ GHz located at the bottom of the first PBG. The simulation result is shown in Fig. 6 (a), where the field pattern is almost symmetric, no SAR effect can be observed. Presumably, the Chern numbers of the photonic bands nearby are zero, only a tiny circulation of the energy flow occurs so that no significant asymmetry reflection is observed. The result is consistent with that in Wang and coworkers' research where they present the Chern number corresponding to the photonic bands (Wang et al., 2008). Similarly, for the upmost PBG ranging from 17 GHz to 19.5 GHz, the SAR effect can not be observed either due to the same reason. Between these two PBGs, there exist another narrow PBG as denoted by a yellow stripe in Fig. 5 (a). We select a frequency $f = 15.5$ GHz as the working frequency to examine the reflection behavior. For convenience, we have given in Fig. 5 (b) the amplified view of the photonic band diagram around this PBG. The corresponding electric field pattern is shown in Fig. 6 (b) where the electric field on the RHS is nearly vanished near the interface, while on the LHS the electric field can be supported so that the SAR effect comes into appearance. The PBG comes from the degeneracy lift resulting from the gyrotropic anisotropy of the MMs. At this frequency range, the energy circulation can be excited, which we will discuss later on.

For comparison, we have also performed the simulation to examine the reflection behavior of a line source from an MPC slab with the same parameters in Wang and coworkers' research (Wang et al., 2008). Concretely, the lattice constant is $a = 38.7$ mm, the radius of the ferrite rod is $r = 0.11a = 4.3$ mm, the corresponding matrix elements of the magnetic permeability are $\mu = 14$ and $\mu' = -12.4$, the working frequency is $f = 4.28$ GHz, and the line source is located $0.25a$ away from the interface. The electric field pattern is shown in Fig. 6 (c) where the EM wave is reflected somewhat leftwards. A weak asymmetry reflection can be observed. In this case, the Chern numbers of the photonic bands around the PBG are not zero as shown by Wang and coworkers (Wang et al., 2008). However, different from the results shown in Fig.

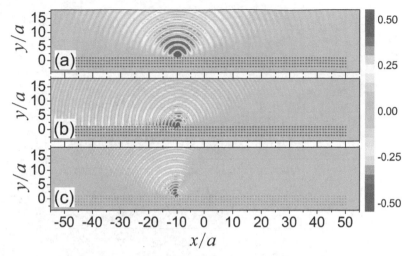

Fig. 6. The reflection behavior for the working frequency located at $f = 9.5$ GHz (a) and $f = 15.5$ GHz (b), corresponding to those in the lower and higher PBGs, respectively, due to the Bragg scattering. The reflection behavior of a line source from an MPC slab with the same parameters in (Wang et al., 2008) is also examined as shown in panel (c).

2, there appears no enhancement of EM field near the interface due to the absence of MSP resonance.

It should be noted that the direction of EM energy circulation does not depend solely on the magnetization, as can be illustrated by comparing Fig. 2 (a) to Fig. 6 (b). To determine the direction of the circulation (clockwise or anticlockwise) we need to calculate the energy circulation around the ferrite rods. For a single rod, it can be evaluated according to

$$\oint \boldsymbol{P} \cdot d\boldsymbol{l} = A(\omega) \sum_n n \left| \frac{b_n}{q_n} \frac{H_n^{(1)}(x)}{J_n(x)} - 1 \right|^2 |q_n J_n(x)|^2, \tag{15}$$

where $A(\omega) = \frac{\pi |E_0|^2}{\mu_b \omega}$ is the prefactor with E_0 the amplitude of the incident wave and μ_b the magnetic permeability of the background medium, $x = k_b r_s$ is the size parameter, b_n and q_n are, respectively, the expansion coefficients of the scattered and incident wave corresponding to the angular momentum n, $J_n(x)$ is n−th order the Bessel function, and $H_n^{(1)}(x)$ is the n−th order Hankel function of the first kind. From Eq. (15), it can be found that the energy circulation depends on the difference of the scattering amplitude for the $|n|$ and the $-|n|$ terms. While the scattering amplitude corresponding to $n = 0$ is not involved. We have calculated the energy circulation around a typical ferrite rod in the first layer of the MM slab. The results show that the circulation is about $-10^{-2} A(\omega)$ for Fig. 2 (a), suggesting a clockwise energy flow, thus explaining the rightwards reflection. For Fig. 6 (a), the energy circulation is nearly zero, corresponding the nearly symmetric reflection. For Fig. 6 (b), the energy circulation is about $10^{-2} A(\omega)$, indicating that an anticlockwise energy flow is formed, explaining the leftwards reflection. While for the case in Fig. 6 (c), the energy circulation is about $10^{-4} A(\omega)$, much weaker than the case in Fig. 2 (a) and Fig. 6 (b). Accordingly, although the leftward reflection occurs, along the interface nearly no EM mode is supported.

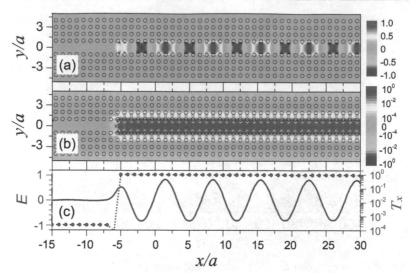

Fig. 7. The profile of the electric field (a) and the x-component of the Poynting vector (b) when a line source of $f = 5$ GHz is located at $(-5.5a, 0)$, between two MM slabs with opposite magnetization and the channel width $D = 2a$. Here $a = 8$ mm is the lattice constant of a square lattice of MM. Panel (c) displays the electric field E at $y = 0$ and the rightward transmitted power T_x versus x as denoted by blue solid line and red dotted line, respectively.

5. Design of an OEMW based on the MSP

Based on the SAR effect we can explore some possible applications. A typical example is due to the remarkable asymmetry of the Poynting vector as shown in Fig. 2 (c). Because of the existence of the PBGs around the MSP resonance, the EM wave can be confined between two MM slabs and transported only in the channel, similar to the conventional photonic crystal waveguide. If two MM slabs have the opposite magnetization, then the EM wave reflected forward from one MM slab will also be reflected forward from the other, leading to the design of an OEMW different from those proposed recently (Fu et al., 2011; 2010; He et al., 2010; Huang & Jiang, 2009; Wang et al., 2008; 2009; Yu et al., 2008; Zhu & Jiang, 2010). The performance of the OEMW is illustrated in Fig. 7 (a) where a line source is placed at $(-5.5a, 0)$, in the middle of two MM slabs with the channel width $D = 2a$. The parameters for the MM slabs and the operating frequency of the line source are the same as in Fig. 2. It can be observed that the EM wave propagates rightward as demonstrated by the x component P_x of the Poynting vector $\boldsymbol{P}$ shown in Fig. 7 (b). To further illustrate the one-way characteristic, we display in Fig. 7 (c) the electric field along $y = 0$ and the rightward transmitted power, $T_x = \int_{-3a}^{3a} P_x dy$, as the functions of the position x. In addition, the EM energy exhibits an extremely low decay rate that is less than 0.1 dB/λ when taking into account of the realistic material absorption (Pozar, 2004). To search out the physical essence, in the following, we will neglect the damping for simplicity.

5.1 Channel width dependence of the OEMW

As already shown in Fig. 4, the SAR effect is dependent on the source-interface separation and it is only workable when the separation is small. This suggests that the OEMW designed

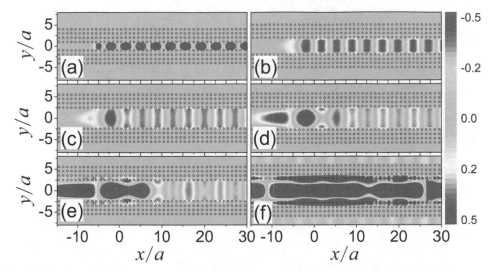

Fig. 8. The electric field patterns for the OEMW operated with different channel widths. Panels (a), (b), (c), (d), (e), and (f) correspond to the cases for the channel widths $D = 2.0a$, $D = 4.0a$, $D = 5.5a$, $D = 5.6a$, $D = 5.7a$, and $D = 6.0a$, respectively.

according to this effect should also depend on the channel width. Physically, as the channel width becomes larger than a wavelength, one part of the EM wave can propagate along the channel without experiencing any of the reflection. Therefore, the unidirectional propagation characteristic shown in Fig. 7 will be diminished. In Fig. 8, we have presented the electric field patterns for the OEMW with different channel widths. Panels (a), (b), (d), (e), and (f) correspond to channel widths $D = 2.0a$, $D = 4.0a$, $D = 5.5a$, $D = 5.6a$, $D = 5.7a$, and $D = 6.0a$, respectively. For OEMW with the channel width $D < 5.5a$, the leftward propagating EM wave is suppressed completely. The OEMW can be considered as a unidirectional device with good performance. However, with the increase of the channel width more and more EM field leaks leftwards as can be observed in Figs. 8 (d)-(f). This result is in good agreement with the result shown in Fig. 4 in that the unidirectionality becomes weaker and weaker with the increase of the source-interface separation. When the channel width is increased to $D = 6.0a$, nearly 0.8λ, no obvious difference can be observed for the leftward and rightward propagating EM waves.

5.2 Robustness against defect, disorder, and inhomogeneity of EMF

A particular important issue of current interest is the robustness of a designed EM devices against defect and disorder. We have performed the simulations to illustrate the issue, which indicates that the one-way waveguiding property based on the SAR effect appears to be immune to defect and disorders, as demonstrated typically in Figs. 9 (b-d). Figure 9 (b) simulates the electric field pattern when a finite linear array of close-packed perfect electrical conductor (PEC) rods is inserted to block the channel. The radius of the PEC rod is $r_p = \frac{1}{2}r_s$. The linear array ranges from $y = -2.5a$ to $y = 2.5a$, forming a drastic defect extending over $\frac{2}{3}\lambda$ in length. The EM wave is seen to circumvent the PEC defect, maintaining a nearly complete power transmission T_x along the channel, as shown in Fig. 9 (j). In Fig. 10, we present a

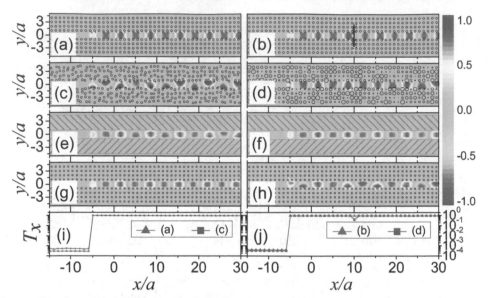

Fig. 9. The electric field patterns for the OEMW operated under the conditions (a) the same as Fig. 7 (a), given here for comparison, (b) with the introduction of a drastic PEC defect to block the channel, (c) with position disorder, and (d) with size fluctuation. The averaged electric field patterns for 20 configurations with position disorder (e) and size fluctuation (f) are also given. The inhomogeneity of the EMF is considered as well for the cases of (g) inhomogeneous and symmetric EMF, and (h) inhomogeneous and asymmetric EMF. In addition, the normalized transmitted power versus x for (a), (c) and (b), (d) are plotted and presented in panels (i) and (j), respectively.

simple schematic diagram to show how the OEMW works and how the EM wave circumvents the PEC defect, which could be helpful for the understanding of mechanism dominating the behavior. For the perfect case without any defect, the MSP resonance can induce the formation of a unidirectional energy circulation so that the channel can only support the energy flow in one direction, the OEMW is thus designed as shown in Fig. 10 (a). When a PEC defect is inserted into the channel, an equivalent waveguiding channel can be created between the PEC defect and the MM as marked by the dashed vertical arrows. Therefore, the EM wave can get around the PEC defect without experiencing any backscattering by propagating along this equivalent channel, resulting in a nearly complete energy transmission. Compared with the defect free waveguide shown in Fig. 9 (a), it can be observed that the defect changes only the phase of the rightward propagating wave, as a result of the delay due to the PEC defect.

The designed OEMW is also robust against the position disorder and the size fluctuation of the ferrite rods as illustrated in Figs. 9 (c) and 9 (d) where we present the corresponding electric field patterns when these two types of perturbations are involved. Position disorder is controlled by a series of random numbers, which introduce a maximal coordinate variation equal to $\frac{1}{4}a$. The size fluctuation of the rod radius is uniform up to 50% of the unperturbed case. It can be observed that the disorder only alters the field patterns, but not the power transmission through the channel as shown in Figs. 9 (i) and (j). Actually, for even stronger perturbation the OEMW still works well provided that the channel is not destroyed. For

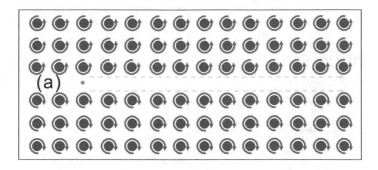

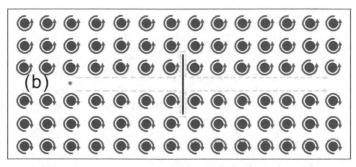

Fig. 10. The simple schematic picture to illustrate how the OEMW works (a), and how the EM wave can get around the defect without any back scattering (b) where the point denote the line source, the circular arrows denote the energy circulation supported by the ferrite rods, and the solid black line denote the PEC defect.

the edge state waveguide, (Wang et al., 2008) the working frequency lies in the Bragg type PBG, so although the system can be immune to the defect, it may suffer from disorder of the building blocks. In Figs. 9 (e) and (f), we also present the averaged electric field patterns over 20 configurations of position and radius disordered systems, respectively. It can be found that the averaged field patterns bear resemblance to that without any perturbation, owing to cancellation of the field at the irregular interface for different configurations. The cases corresponding to different amplitude of perturbation are examined as well, similar behaviors can be observed, indicating the statistical validity of the results.

In realistic situation, the EMF can not be perfectly homogeneous. For this reason, we also present the simulation results for the OEMW operated under an inhomogeneous EMF. Figure 9 (g) corresponds to the case when a symmetric inhomogeneous EMF is exerted. The EMFs at the different layers from inside to outside are, respectively, 890 Oe, 910 Oe, 930 Oe, and 950 Oe, symmetrically for the upper and lower MM slabs. The situation corresponding to the asymmetric inhomogeneous EMF is also considered as shown in Fig. 9 (h) where the EMFs at different layers from inside to outside are 870 Oe, 900 Oe, 910 Oe, 950 Oe for the upper MM slab and 920 Oe, 930 Oe, 940 Oe, 970 Oe for the lower MM slab. It can be observed that the symmetric distribution of the EMF leads to a symmetric electric field pattern and the asymmetric EMF can shape the electric field pattern into an asymmetric one. However, a good performance can still be maintained. This is also a favorable aspect for the design of the EM devices.

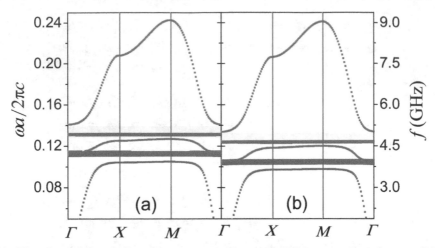

Fig. 11. The photonic band diagrams corresponding to the MM operated under two different EMFs (a) $H_0 = 900$ Oe and (b) $H_0 = 800$ Oe, respectively. The other parameters of the MM are the same as those used in Fig. 1. The working frequency ranges for the OEMW under different EMFs are denoted with the yellow stripes.

5.3 Tunability of the working frequency

A special property of the OEMW designed in our work is the tunability of the working frequency due to the dependence of the MSP resonance on the EMF. This tunability can be clearly seen by examining the photonic band diagrams of the MM under different EMFs as shown in Figs. 11 (a) and (b), corresponding to $H_0 = 900$ Oe and $H_0 = 800$ Oe, respectively. The yellow stripes mark the working frequency range of the OEMW, the flat bands there correspond to the MSP resonance. When the EMF decreases from $H_0 = 900$ Oe to $H_0 = 800$ Oe, the flat bands move downwards, so does the working frequency. In this manner, the working frequency can be controlled by an EMF. Due to the sensitivity of the MSP resonance dependent on the EMF, the working frequency of the OEMW can be manipulated easily.

To give a clear picture how the OEMW works under different EMF, we present in Fig. 12 the electric field patterns of the OEMW under $H_0 = 900$ Oe [(a), (b)] and $H_0 = 800$ Oe [(c), (d)], respectively. The other parameters of the OEMW are the same as those in Fig. 7. In panel (a), the frequency is $f = 5.2$ GHz, lying in the working frequency range. As can be observed, the EM field is confined in the channel and a good one-way propagating behavior is manifested. When the frequency is decreased to $f = 4.7$ GHz under the same EMF, it lies outside the working frequency range, the EM field leaks outside the channel as demonstrated in Fig. 12 (b). Accordingly, the EM field propagating along the channel becomes weaker and weaker so that at this frequency it can not be operated as an OEMW. Nonetheless, by tuning EMF to a lower value $H_0 = 800$ Oe, $f = 4.7$ GHz is located in the new working frequency range as can be confirmed from the photonic band diagram in Fig. 11 (b). The corresponding electric field pattern is shown in Fig. 12 (c) where we can observe that the OEMW works very well. Similar to the case shown in Fig. 12 (b), if the frequency is tuned to $f = 5.2$ GHz, lying outside the working range under $H_0 = 800$ Oe. Then, the device can not be operated as an OEMW by examining the electric field pattern shown in Fig. 12 (d) where we can see that most of the EM field leaks outside the channel. From the above analysis, we can conclude that by tuning the

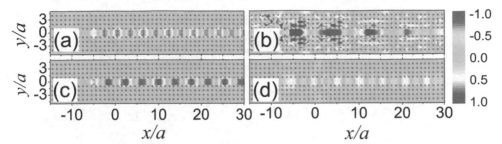

Fig. 12. The electric field patterns showing the OEMW working at frequency $f = 5.2$ GHz (a), (d) and $f = 4.7$ GHz (b), (c). The exerted EMFs are $H_0 = 900$ Oe and $H_0 = 800$ Oe for (a)-(b) and (c)-(d), respectively. All the other parameters are the same as those in Fig. 7.

EMF from $H_0 = 900$ Oe to $H_0 = 800$ Oe, the working frequency can be adjusted from 5.2 GHz to 4.7 GHz. Accordingly, by tuning the EMF we can manipulate the working frequency of the OEMW. However, for the OEMW designed based solely on the TRS breaking, the working frequency is inert to the EMF so that no manipulability can be expected.

6. Design of beam bender and splitter

Based upon the SAR effect, we can also design other EM waveguiding devices by constructing a corner configuration. Typical simulation results of such applications are presented in Fig. 13 where the configuration of the system is similar to that of the OEMW except that a cladding slab is added on top of the system. For a line source located in the vertical channel at $(0, -15a)$, the EM wave is seen to propagate upward first, operated in the same manner as an OEMW. After that, the EM waves make a $90°$ turn at the corner without any backward scattering so that nearly 100% power transmission is realized for the beam bender, as shown quantitatively in Fig. 13 (e) by the blue solid line. The power transmission rates are defined as T_x/T_y with T_x and T_y the EM energy propagating along the two different parts of the channel in the x (rightward) and the y (upward) directions. They are calculated numerically according to $T_x = \int_{-3a}^{3a} P_x \mathrm{d}y$ and $T_y = \int_{-3a}^{3a} P_y \mathrm{d}x$. By reversing the magnetization of the ferrite rods with the coordinates $x < 0$ in the upper cladding slab, the EM wave can be divided equally into two branches at the bifurcation point $x = 0$ as shown in Fig. 13 (b). In each wing the EM wave is transported with 50% power transmission in some frequency range, as is shown in Fig. 13 (e) by the red dashed line. In the frequency range $f > 5.04$ GHz the power transmission rate for the beam bender decreases obviously as can be observed in Fig. 13 (e). To explain this, we have presented the electric field patterns for the beam bender and splitter operated at the frequency $f = 5.1$ GHz. The results are shown in Figs. 13 (c) and (d), respectively. For the beam bender, due to the difference of the light path at the upper and lower interfaces the wave front is severely distorted. In addition, as the frequency deviates from the MSP resonance, the energy circulation around the ferrite rods becomes weaker. Therefore, some part of the EM wave leaks leftwards as shown in Fig. 13 (c), resulting in the drop of the transmission rate. Since the orientation of the magnetization can be controlled by an EMF, the function of the system can be switched between bender and splitter. This makes the device more flexible and favorable in practical applications.

The unidirectional waveguiding device designed in our work can still be operable with a good performance in a deep subwavelength scale. In Fig. 14, we present the simulation results. The

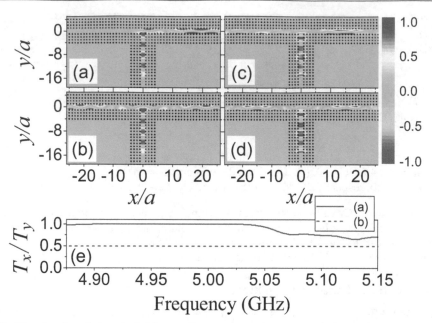

Fig. 13. The simulated electric field patterns corresponding to a $90°$ shaper beam bender (a) and a beam splitter (b) operating at $f = 5$ GHz under the EMF $H_0 = 900$ Oe. The electric field patterns for the beam bender and splitter working at $f = 5.1$ GHz are also presented in panels (c) and (d), respectively. The lattice constant is $a = 8$ mm, kept unchanged and the line source is located at $(0, -15a)$. The curves in (e) give the transmissivity T_x/T_y plotted as the functions of frequency.

waveguiding device considered in panels (a), (b), and (c) are designed in the same manner as those in Figs. 7 (a), 13 (a), and 13 (b), respectively, except that the MM is scaled down in size, with $a = 2$ mm and $r = \frac{1}{4}a$, while keeping the working frequency $f = 5$ GHz unchanged. In such situation, the working wavelength $\lambda = 60$ mm is nearly 30 times the lattice constant a. It can be observed from the field patterns that a superior subwavelength confining and steering is realized in a straight OEMW, a sharp beam bender, and a beam splitter. The numerical calculations indicate that the full lateral width at half maximum field intensity $w_h < 0.1\lambda$. The power transmissivity is also simulated for the beam bender and splitter as shown in Fig. 14 (d) where the red dashed line and the blue solid line correspond to panels (b) and (c), respectively. It can be seen that the waveguiding devices still exhibit a high transmission efficiency as well as a finite band width. In addition, compared with the results shown in Fig. 13 (e), it can be seen that a shift of the working frequency is demonstrated, which originates from the enhancement of the coupling strength due to the decrease of the lattice separation. Comparing with the results shown in Fig. 13, we can also find that the transmission efficiency for the beam bender is much improved. The reason lies in that the decrease of the channel width will ease up the distortion of the EM field, which is clearly demonstrated in Fig. 14 (b). Accordingly, the power leakage of the EM wave to the left wing of the horizonal channel can be decreased. Besides, just like the case shown in Fig. 13 the function of the system is still magnetically switchable between beam bender and splitter.

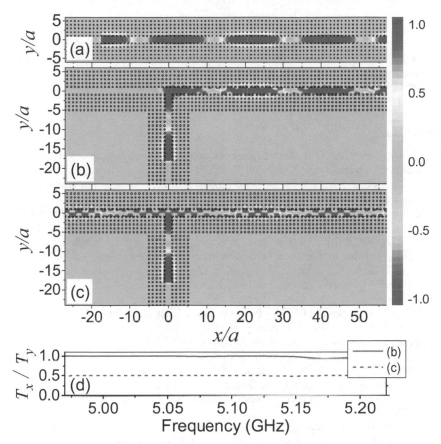

Fig. 14. The electric field patterns showing the device operable at the deep subwavelength scale. Panels (a), (b), and (c) are the same as Fig. 7(a), Fig. 13(a), and Fig. 13(b), respectively, except that the lattice constant of the MM is $a = 2$ mm and rod radius $r = \frac{1}{4}a$. The curves in (d) gives the rightward transmissivity for the beam bender and beam splitter.

7. Conclusion

In summary, we have demonstrated a very interesting SAR effect occurring at the interface of an MM slab. The Mie theory and the multiple scattering theory used in our simulation have also been introduced. Our results suggest that the SAR effect originates from combined action of MSP resonance and TRS breaking nature of MM under an EMF. We also examine the dependence of SAR effect on the frequency and the source-interface separation. Possible applications have been demonstrated by designing a straight OEMW, a sharp beam bender, and a beam splitter. An excellent performance of the device can be achieved. In particular, our design can even be operable in the deep subwavelength scale. Actually, a lot more issues can be expected with appropriate design of the MM such as the muti-channel unidirectional filter, cloaking, and also the zero index materials.

8. Acknowledgements

This work is supported by the China 973 program, NNSFC (10774028, 10904020), MOE of China (B06011), SSTC (08dj1400302), China postdoctoral science foundation (200902211), the open project of SKLSP in Fudan Univeristy (KL2011_8), Zhejiang Provincial Natural Science Foundation of China (Y12A040009) and Zhejing Normal University initiative foundation. STC is partly supported by the DOE.

9. References

Ao, X. Y.; Lin, Z. F. & Chan, C. T. (2009). One-way edge mode in a magneto-optical honeycomb photonic crystal. *Phys. Rev. B*, Vol. 80, No. 3, 033105

Barnes, W. L.; Dereux, A.; Ebbesen, T. W. (2003). Surface plasmon subwavelength optics. *Nature*, Vol. 424, No. 6950, 824-830

Bohren, C. F. & Huffman, D. R. (1983). *Absorption and Scattering of Light by Small Particles*, Wiley, New York.

Burgos, S. P.; Waele, R. de; Polman, A. & Atwater, H. A. (2010). A single-layer wide-angle negative-index metamaterial at visible frequencies. *Nat. Mater.* Vol. 9, No. 5, 407-412

Chen, P.; Wu, R. X.; Xu J.; Jiang, A. M. & Ji, X. Y. (2007). Effect of magnetic anisotropy on the stop band of ferromagnetic band gap materials. *J. Phys.: Condens. Matter*, Vol. 19, No. 10, 106205

Chew, W. C. (1995). *Waves and Fields in Inhomogenous Media*, IEEE Press, New York.

Chui, S. T. & Lin, Z. F. (2007). Probing states with macroscopic circulations in magnetic photonic crystals. *J. Phys.: Condens. Matter*, Vol. 19, No. 40, 406233

Chui, S. T.; Liu, S. Y. & Lin Z. F. (2010). Reflected wave of finite circulation from magnetic photonic crystals. *J. Phys.: Condens. Matter*, Vol. 22, No. 18, 182201

Dolling, G.; Enkrich, C.; Wegener, M.; Soukoulis, C. M. & Linden, S. (2006). Simultaneous negative phase and group velocity of light in a metamaterial. *Science*, Vol. 312, No. 5775, 892-894

Eggimann, W. H. (1960). Scattering of a plane wave on a ferrite cylinder at normal incidence. *IRE Trans. Microwave Theory and Tech.*, Vol. 8, No. 4, 440-445

Engheta, N. (2007). Circuits with light at nanoscales: Optical nanocircuits inspired by metamaterials. *Science*, Vol. 317, No. 5845, 1698-1702

Ergin, T.; Stenger, N.; Brenner, P.; Pendry, J. B. & Wegener, M. (2010). Three-dimensional invisibility cloak at optical wavelengths. *Science*, Vol. 328, No. 5976, 337-339

Fang, N.; Lee, H.; Sun, C. & Zhang, X. (2005). Sub-diffraction-limited optical imaging with a silver superlens. *Science*, Vol. 308, No. 5721, 534-537

Felbacq, D.; Tayeb, G. & Maystre, D. (1994). Scattering by a random set of parallel cylinders. *J. Opt. Soc. Am. A*, Vol. 11, No. 9, 2526-2538

Fu, J. X.; Lian, J.; Liu, R. J.; Guan, L. & Li, Z. Y. (2011). Unidirectional channel-drop filter by one-way gyromagnetic phtonic crystal waveguides. *Appl. Phys. Lett.*, Vol. 98, No. 21, 211104

Fu, J. X.; Liu, R. J.; & Li, Z. Y. (2011). Robust one-way modes in gyromagnetic photonic crystal waveguides with different interfaces. *Appl. Phys. Lett.*, Vol. 97, No. 4, 041112

Giannini,V.; Vecchi, G. & Rivas, J. G. (2010). Lighting up multipolar surface plasmon polaritons by collective resonances in arrays of nanoantennas. *Phys. Rev. Lett.*, Vol. 105, No.26, 266801

Gollub, J. R.; Smith, D. R.; Vier, D. C.; Perram, T.; Mock, J. J. (2005). Experimental characterization of magnetic surface plasmons on metamaterials with negative permeability. *Phys. Rev. B*, Vol. 71, No. 19, 195402

Haldane, F. D. M. & Raghu, S. (2008). Possible realization of directional optical waveguides in photonic crystals with broken time-reversal symmetry. *Phys. Rev. Lett.*, Vol. 100, No. 1, 013904

He, C.; Chen, X. L.; Lu, M. H.; Li, X. F.; Wan, W. W.; Qian, X. S.; Yin, R. C. & Chen, Y. F. (2010). Tunable one-way cross-waveguide splitter based on gyromagnetic phtonic crystal. *Appl. Phys. Lett.*, Vol. 96, No. 11, 111111

Huang, C. & Jiang, C. (2009). Nonreciprocal photonic crystal delay waveguide. *J. Opt. Soc. Am. B*, Vol. 26, No. 10, 1954-1958

Jin, J. J.; Liu, S. Y.; Lin, Z. F. & Chui, S. T. (2009). Effective-medium theory for anisotropic magnetic metamaterials. *Phys. Rev. B*, Vol. 80, No. 11, 115101

Leonhardt, U. (2006). Optical conformal mapping. *Science*, Vol. 312, No. 5781, 1777-1780

Lezec, H. J.; Dionne, J. A. & Atwater, H. A. (2007). Negative refraction at visible frequencies. *Science*, Vol. 316, No. 5823, 430-432

Lai, Y.; Chen, H. Y.; Zhang, Z. Q. & Chan, C. T. (2009). Complementary media invisibility cloak that cloaks objects at a distance outside the cloaking shell. *Phys. Rev. Lett.*, Vol. 102, No. 9, 093901

Leung, K. M. & Qiu, Y (1993). Multiple-scattering calculation of the two-dimensional phtonic band structure. *Phys. Rev. B*, Vol. 48, No. 11, 7767-7771

Li, J. & Pendry, J. B. (2008). Hiding under the carpet: A new strategy for cloaking. *Phys, Rev. Lett.*, Vol. 101, No. 20, 203901

Liu S. Y. & Lin, Z. F. (2006). Opening up complete photonic bandgaps in three-dimensional photonic crystals consisting of biaxial dielectric spheres. *Phys. Rev. E*, Vol. 73, No. 6, 066609

Liu, Z. W.; Lee, H.; Xiong, Y.; Sun, C. & Zhang, X. (2007). Far-field optical hyperlens magnifying sub-diffraction-limited objects. *Science*, Vol. 315, No. 5819, 1686

Liu, S. Y.; Chen, W. K.; Du, J. J.; Lin, Z. F.; Chui, S. T. & Chan, C. T. (2008). Manipulating negative-refractive behavior with a magnetic field. *Phys. Rev. Lett.*, Vol. 101, No. 15, 157407

Liu, S. Y.; Du, J. J.; Lin, Z. F.; Wu, R. X. & Chui S. T. (2008). Formation of robust and completely tunable resonant photonic band gaps. *Phys. Rev. B*, Vol. 78, No. 15, 155101

Liu, R.; Ji, P. C.; Mock, J.; Chin, J. J. Y.; Cui, T. J. & Smith, D. R. (2009). Broadband ground-plane cloak. *Science*, Vol. 323, No. 5912, 366-369

Liu, S. Y.; Lu, W. L.; Lin, Z. F. & Chui, S. T. (2010). Magnetically controllable unidirectional electromagnetic waveguiding devices designed with metamaterials. *Appl. Phys. Lett.*, Vol. 97, No. 20, 201113

Liu, S.Y.; Lu, W. L.; Lin, Z. F. & Chui, S. T. (2011). Molding reflection from metamaterials based on magnetic surface plasmons. *Phys. Rev. B*, Vol. 84, No.4, 045425

Noginov, M. A.; Zhu, G.; Mayy, M.; Ritzo, B. A.; Noginova, N. & Podolskiy, V. A. (2008). Stimulated emission of surface plasmon polaritons. *Phys. Rev. Lett.*, Vol. 101, No. 22, 226806

Ozbay, E.(2006). Plasmonics: Merging photonics and electronics at nanoscale dimensions. *Science*, Vol. 311, No. 5758, 189-193

Pendry, J. B. (2000). Negative refraction makes a perfect lens. *Phys. Rev. Lett.*, Vol. 85, No. 18, 3966-3969

Pendry, J. B.; Holden, A. J.; Stewart, W. J. & Youngs, I. (1996). Extremely low frequency plasmons in metallic mesostructures. *Phys. Rev. Lett.*, Vol. 76, No. 25, 4773-4776

Pendry, J. B.; Holden, A. J.; Holden, D. J. & Stewart, W. J. (1999). Magnetism from conductors and enhanced nonlinear phenomena. *IEEE Tran. Microwave Theory Tech.*, Vol. 47, No. 11, 2075-2084

Pendry, J. B.; Schurig, D. & Smith, D. R. (2006). Controlling electromagnetic fields. *Science*, Vol. 312, No. 5781, 1780-1782

Pozar, D. M. (2004). *Microwave Engineering*, Wiley, New York.

Poo, Y.; Wu, R. X.; Lin, Z. F.; Yang, Y. & Chan, C. T. (2011). Experimental realization of self-guiding unidirectional electromagnetic edge states. *Phys. Rev. Lett.*, Vol. 106, No. 9, 093903

Schurig, D.; Mock, J. J.; Justice, B. J.; Cummer, S. A.; Pendry, J. B.; Starr, A. F. & Smith, D. R. (2006). Metamaterial electromagnetic cloak at microwave frequencies. *Science*, Vol. 314, No. 5801, 977-980

Shalaev, V. M. (2007). Optical negative-index metamaterials. *Nature Photon.*, Vol. 1, No. 1, 41-48

Shelby, R. A.; Smith, D. R. & Schultz S. (2001). Experimental verification of a negative index of reflection. *Science*, Vol. 292, No. 5514, 77-79

Slichter, C. P. (1978) *Principle of Magnetic Resonance*, Springer, Berlin.

Smith, D. R.; Padilla, W. J.; Vier, D. C.; Nemat-Nasser, S. C. & Schultz, S. (2000). Composite medium with simultaneously negative permeability and permittivity. *Phys. Rev. Lett.*, Vol. 84, No. 18, 4184-4187

Taubner, T.; Korobkin, D.; Urzhumov, Y.; Shvets, G. & Hillenbrand, R. (2006). Near-field microscopy through a SiC superlens. *Science*, Vol. 313, No. 5793, 1595

Valentine, J.; Zhang, S.; Zentgraf, T.; Ulin-Avila, E.; Genov, D. A.; Bartal, G. & Zhang, X. (2008). Three-dimensional optical metamaterial with a negative refractive index. *Nature*, Vol. 455, No. 7211, 376-379

Veselago, V. G. (1968). The electrodynamics of substrates whith simultaneously negative values of ε and μ. *Sov. Phys. Usp.*, Vol. 10, No. 4, 509-514

Wang, X. D.; Zhang, X. G.; Yu, Q. L. & Harmon, B. N. (1993). Multiple-scattering theory for electromagnetic waves. *Phys. Rev. B*, Vol. 47, No. 8, 4161-4167

Wang, Z.; Chong, Y. D.; Joannopoulos, J. D. & Soljačić, M. (2008). Reflection-free one-way edge modes in a gyromagnetic photonic crystal. *Phys. Rev. Lett.*, Vol. 100, No. 1, 013905

Wang, Z.; Chong, Y. D.; Joannopoulos, J. D. & Soljačić, M. (2009). Observation of unidirectional backscattering-immune topological electromagnetic states. *Nature*, Vol. 461, No. 7265, 772-775

Yu, Z. F.; Veronis, G.; Wang, Z. & Fan, S. H. (2008). One-Way electromagnetic waveguide formed at the interface between a plasmonic metal under a static magnetic field and a photonic crystal. *Phys. Rev. Lett.*, Vol. 100, No. 2, 023902

Zayats, A. V.; Smolyaninov, I. I. & Maradudin, A. A. (2005). Nano-optics of surface plasmon polaritons. *Phys. Rep.*, Vol. 408, No. 3-4, 131-314

Zentgraf, T.; Liu, Y. M.; Mikkelsen, M. H.; Valentine, J. & Zhang, X. (2011). Plasmonic Luneburg and Eaton lenses. *Nat. Nanotechnology*, Vol. 6, No. 3, 151-155

Zhang, S.; Fan, W.; Panoiu, N. C.; Malloy, K. J.; Osgood, R. M. & Brueck, S. R. J. (2005). Experimental demonstration of near-infrared negative-index metamaterials. *Phys. Rev. Lett.*, Vol. 95, No. 13, 137404

Zhu, H. B. & Jiang, C. (2010). Broadband unidirectional electromagnetic mode at interface of anti-parallel magnetized media. *Opt. Express*, Vol. 18, No. 7, 6914-6921

Zia, R.; Schuller, J. A.; Chandran, A. & Brongersma, M. L. (2006). Plasmonics: the next chip-scale technology. *Materials today*, Vol. 9, No. 7-8, 20-27

8

Applications of Artificial Magnetic Conductors in Monopole and Dipole Antennas

Amir Jafargholi[1], Mahmood Rafaei Booket[2] and Mehdi Veysi[1]

[1]*K.N. Toosi University of Technology*
[2]*Tarbiat Modares University*
Iran

1. Introduction

In recent years, introducing metamaterials based on theory established by Veselago, opened the way for many researcher groups to enhance the antenna performances (Veselago, 1968; Veysi et al., 2010; Rafaei et al., 2010; Veysi et al., 2011; Rafaei et al., 2011; Jafargholi et al., 2010; Jafargholi et al., 2011). A standard procedure has been also established for the design of bulk artificial media with negative macroscopic permeability and permittivity. In the past few years, extensive research has been carried out on the metamaterial realization of the artificial magnetic conductors (AMCs) (Jafargholi et al., 2010; McVay et al., 2004). As revealed in (Erentok et al., 2005), the artificial magnetic conductor can be also realized using a volumetric metamaterial constructed from a periodic arrangement of the capacitively loaded loop (CLL) elements. Recently, it has found interesting applications in antenna engineering (Ziolkowski et al., 2003). Due to unique electromagnetic properties, metamaterials have been widely considered in monopole and dipole antennas to improve their performance (Ziolkowski et al., 2003; Liu et al., 2009; Rogers et al., 2003).

This chapter is mainly focused on two different applications of artificial magnetic conductors in the field of dipole pattern modification and dual and multiband dipole antennas. At first, the radiation patterns of the monopole and dipole antennas have been considered, especially at the second harmonic of the main resonant frequency. A closer examination on the PMC structures reveals that they can be used to modify the monopole and dipole radiation patterns. To this aim, the CLLs have been used to realize perfect magnetic conductor.

Finally, this chapter examines reactive loading technique to achieve dual band operation and further size reduction for double-sided printed dipole antennas. Here, the reactive loads are realized by two balanced CLLs placed close to the edge of the printed dipole antenna.

2. Pattern modification of monopole antenna

The linearly polarized and omnidirectional radiation pattern of a vertical monopole antenna has led to a wide range of applications in wireless communications such as wireless local area network (WLAN) and radio broadcast. It is known that the long monopole antennas ($\geq \lambda/2$) suffer from the 180 phase reversal. The fields radiated by reverse current of the long monopole antenna do not reinforce those radiated by original current, resulting in very poor radiation efficiency (Balanis, 1989). As a result, the antenna radiation pattern does not remain omnidirectional within the interested frequency range. In the conventional monopole antenna, the resonance frequencies ω_m correspond to the frequencies where the physical length L of the monopole is an odd multiple of quarter-wavelength. In other words, the antenna resonance frequencies are harmonics of the design frequency ω_1. However, omnidirectional radiation pattern distortion and low directivity are two major disadvantages associated with monopole resonating at higher order harmonics ($\omega_m > \omega_1$). In other words, conventional monopole antenna only radiates an omnidirectional radiation pattern at the design frequency ω_1. In order to have omnidirectional radiation pattern within the antenna bandwidth (ranging from f_L to f_U), the monopole length has to be less than $\lambda_U/2$, where λ_U is the free space wavelength at f_U. However, the antenna directivity decreases because of the significant reduction in the monopole length.

In this section, the use of AMCs to load a monopole antenna has been investigated. It is known that the current reversal, that occurs at frequencies much beyond the antenna natural frequency, disturbs the omnidirectional radiation pattern of the monopole antenna. The current distribution of the monopole antenna can be improved to a large extent by using PMC loading. To this aim, the CLLs are used to realize perfect magnetic conductor behavior.

2.1 PMC loaded monopole antenna

Due to the reverse current effects, the monopole radiation pattern does not remain omnidirectional at the second harmonic of the main resonant frequency (Balanis, 1989). In this section, a monopole antenna loaded with the PMC layer is proposed to increase omnidirectional radiation bandwidth. To make the concept more clear, three ideal models are simulated, all of which are partly covered by a very thin PMC shell, as shown in Fig. 1. As a reference, a conventional monopole antenna is also simulated for comparison. It has the same dimensions as the geometries in Fig. 1, except that the PMC cover is removed.

Fig. 2, shows the simulated reflection coefficient of the monopole antennas with and without the PMC cover. The resonant frequencies for case I and II are 18.5GHz and 24GHz, respectively, whereas the resonant frequency for case III remains the same as the conventional monopole antenna. For the conventional monopole antenna, distortion of the omnidirectional radiation pattern occurs at frequencies higher than 20GHz. This upper limit is indicated by dashed line in Fig. 2, and considered as an antenna length limitation.

Fig. 3, shows the radiation patterns of the monopole antennas with and without PMC cover when frequency varies from 12GHz to 30GHz in 1GHz increments. As revealed in the figure, when the monopole antenna is loaded with the PMC cover (case III), the antenna radiation pattern considerably improves as compared to that of the conventional monopole antenna, especially at the second harmonic of the main resonant frequency (29GHz).

Fig. 4, compares the simulated directivity of the monopole antennas with and without the PMC cover in the azimuth plane. As is evident from Fig. 4, the directivity curve for the case II is approximately flat while the antenna directivity for the case III significantly improves as compared to that of the conventional monopole antenna. For our discussion on the pattern modification, the results shown in Figs. 2 to 4, need to be considered simultaneously.

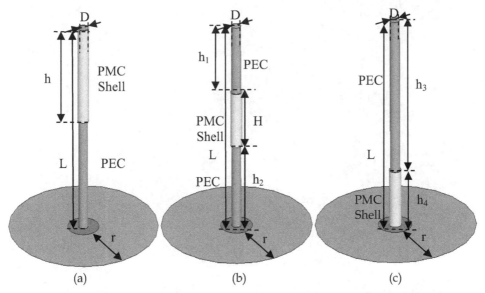

Fig. 1. Three ideal models of PMC loaded monopole antennas, (a) case I, (b) case II, and (c) case III: L=7.5mm, D=0.5mm and h=3.5mm, h_1=2.5mm, h_2=3mm, H=2mm, h_3=5.5mm, h_4=2mm, and r =4mm. From (Jafargholi et al., 2010), copyright © 2010 by the Institute of Electrical and Electronics Engineers (IEEE).

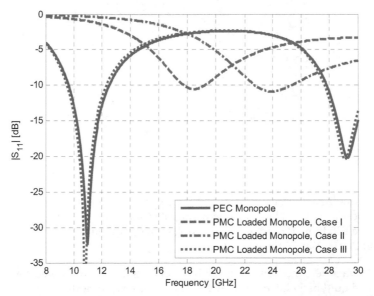

Fig. 2. Reflection coefficient of monopole antennas with and without PMC cover. From (Jafargholi et al., 2010), copyright © 2010 by the Institute of Electrical and Electronics Engineers (IEEE).

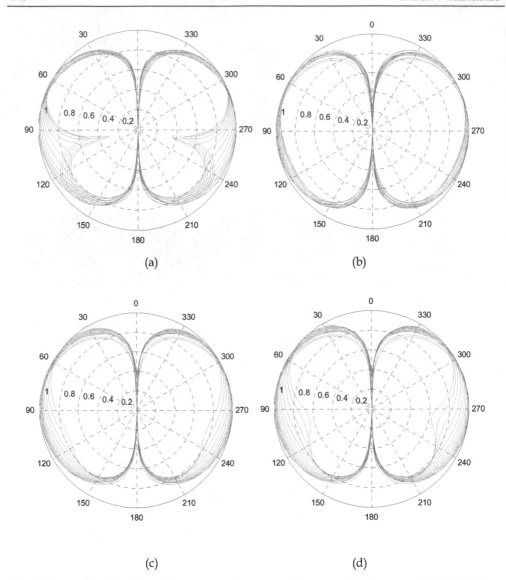

(a)   (b)

(c)   (d)

Fig. 3. Normalized Radiation Patterns v.s. frequency (a) conventional monopole, (b) PMC loaded monopole (case I), (c) PMC loaded monopole (case II) , and (d) PMC loaded monopole (case III). From (Jafargholi et al., 2010), copyright © 2010 by the Institute of Electrical and Electronics Engineers (IEEE).

Consequently, the ideal model shown in Fig. 1 (c) (case III) is considered for the practical realization. It should be pointed out that we can always use a shorter monopole antenna to improve the omnidirectional radiation pattern. However, the prices we pay are the higher resonant frequency and lower directivity due to the significant reduction in the monopole length.

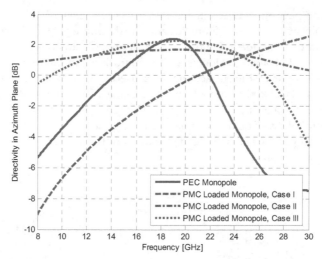

Fig. 4. Simulated directivity (maximum) of monopole antennas with and without PMC cover (on a finite ground plane) in azimuth plane. From (Jafargholi et al., 2010), copyright © 2010 by the Institute of Electrical and Electronics Engineers (IEEE).

2.2 CLL loaded monopole

In the previous section, it was revealed that the suppression of phase reversal by incorporating PMC cover has led to the improved radiation pattern, especially at the second harmonic of the main resonant frequency. In this section, the capacitively loaded loops (CLLs) are used to realize an artificial PMC (Erentok et al., 2005). Fig. 5 shows a CLL loaded monopole antenna together with the basic unit cell of the CLL structure. The finite two-CLL-deep metamaterial AMC cover was designed separately (without the ground plane) to operate at around 26GHz. The CLL dimensions were then optimized to obtain good radiation patterns. The total number of the CLL elements is 44 and the separation between the CLL elements is 0.5mm.

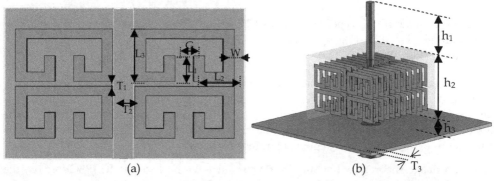

Fig. 5. A schematic view of (a) CLL unit cell: L_1=0.67mm, L_2=0.89mm, L_3=1.37mm, W= 0.247mm, G=0.411mm, T_1= 0.13mm, T_2= 0.5mm, and (b) CLL loaded monopole antenna on a finite ground plane: h_1= 3.25mm, h_2 =4mm, h_3=0.25mm, T_3=0.45mm. From (Jafargholi et al., 2010)], copyright © 2010 by the Institute of Electrical and Electronics Engineers (IEEE).

Fig. 6 shows a comparison between the reflection coefficients of the CLL loaded and unloaded monopole antennas of 7.5mm length. Furthermore, Fig. 7 shows the surface current densities on the CLL loaded and unloaded monopole antennas. It can be seen that the current in the CLL loaded region is the superposition of two currents oriented in opposite directions. However, the surface current caused by CLL cells is dominant, and thus the current in all parts of the monopole has the same phase. To more understand the operation mechanism of the CLL loaded monopole antenna, we assume that the monopole antenna is surrounded by CLL cells, where the current direction of the monopole is reversed. Fig. 8, conceptually explains the distribution of the surface current density on the CLL loaded monopole antenna.

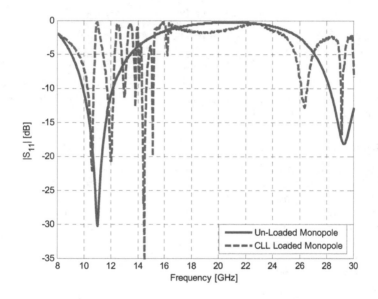

Fig. 6. Simulated reflection coefficient of the CLL loaded and unloaded monopole antennas. From (Jafargholi et al., 2010), copyright © 2010 by the Institute of Electrical and Electronics Engineers (IEEE).

A rigorous explanation must consider the complex interactions between the monopole and the CLL structure, such as the effects of the finite dimensions of the CLL structure on the current distribution. Consequently, full wave analysis methods have to be used in the antenna designs. However, to simplify the analysis, one can assume that the transverse dimensions of the CLL loaded region are infinite in extent.

Based on image theorem, when an electric current is vertical to a PMC (PEC) region, the current image has the reversed (same) direction. For the current at the bottom of the monopole flowing into the CLL loaded region, the CLL loaded region acts as a PMC cover (Erentok et al., 2005). Consequently, the direction of the image current is opposite to that of the original current, as shown in Fig. 8.

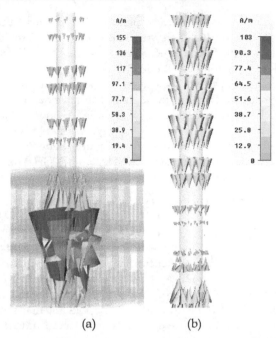

Fig. 7. Surface current densities on the (a) CLL loaded monopole and (b) unloaded monopole at f= 25GHz. From (Jafargholi et al., 2010), copyright © 2010 by the Institute of Electrical and Electronics Engineers (IEEE).

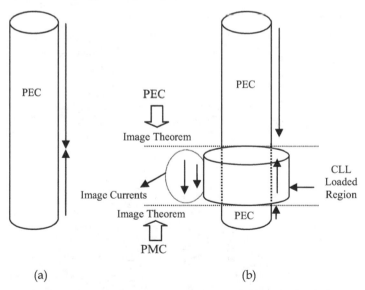

Fig. 8. Conceptual schematics of the (a) conventional and (b) PMC-loaded monopoles. From (Jafargholi et al., 2010)], copyright © 2010 by the Institute of Electrical and Electronics Engineers (IEEE).

In contrast, for the current at the top of monopole flowing into the CLL loaded region, the CLL loaded region acts as an artificial electric conductor (AEC) (Erentok et al., 2005), and thus the image current in the CLL loaded region has the same direction as the original current. The total surface current in the CLL loaded region is obtained as the sum of the two image currents and original current. Consequently, the current phase of the CLL loaded monopole antenna remains unchanged throughout the antenna, as shown in Fig. 8. The radiation patterns of the conventional and CLL loaded monopole antennas on a finite ground plane are shown in Figs. 9 and 10, respectively. As can be seen, when the monopole antenna is loaded with the CLL structure, the radiation patterns improve significantly, especially at the second harmonic (26GHz, 29GHz, and 30GHz) of the main resonant frequency where antenna is matched well.

Also, the antenna radiation efficiency is reasonably high over a wide frequency window, despite the material loss in copper and the CLL metamaterial features. Although, simulation results confirm the modification of the antenna radiation patterns at frequencies up to around 30GHz, the antenna reflection coefficient needs to be modified by impedance matching techniques, especially at frequencies far from the antenna resonant frequencies (Erentok et al., 2008). The simulated gains of the CLL loaded and unloaded monopole antennas in azimuth plane are compared in Fig. 11. As compared to the conventional monopole antenna, the gain of the CLL loaded monopole antenna significantly increases, especially at the higher frequencies. These comparisons demonstrate the unique capability of the AMCs to improve the antenna radiation bandwidth. It is worth noting that the antenna radiation patterns are not completely symmetric in the frequency band from 12GHz to 30GHz because of the asymmetric geometry.

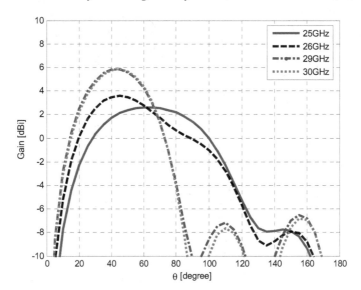

Fig. 9. Radiation patterns of the conventional monopole antenna versus frequency at $\varphi=0°$. From (Jafargholi et al., 2010), copyright © 2010 by the Institute of Electrical and Electronics Engineers (IEEE).

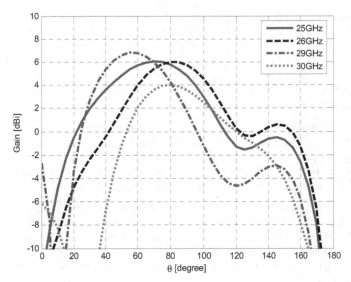

Fig. 10. Radiation patterns of the CLL loaded monopole antenna versus frequency at $\varphi=0^0$. From (Jafargholi et al., 2010), copyright © 2010 by the Institute of Electrical and Electronics Engineers (IEEE).

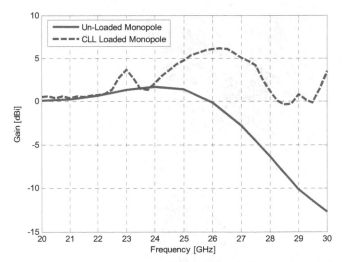

Fig. 11. Comparison of the maximum gains of the monopole antennas with and without CLL covers (on a finite ground plane) in the azimuth plane. From (Jafargholi et al., 2010), copyright © 2010 by the Institute of Electrical and Electronics Engineers (IEEE).

2.3 Pattern modification of dipole antenna

Dipole antennas are preferable in modern wireless communication systems, especially their printed types, due to their low profile, light weight and low fabrication cost as well as their compatibility with microwave and millimeter wave circuits. It was revealed in (Erentok et al., 2007; Erentok et al., 2008) that an efficient and electrically small magnetic based antenna can be realized by adding a planar interdigitated CLL element to a rectangular semi-loop antenna, which is coaxially-fed through a finite ground plane. The performance of a printed dipole antenna near a 3D-CLL block has been also examined in (Zhu et al., 2010). Recently, the use of TL-MTM to load antennas has been investigated in (Antoniades et al., 2009; Zhu et al., 2009). A miniaturized printed dipole loaded with left-handed transmission lines is also proposed in (Iizuka et al., 2006; Iizuka et al., 2007).

However, it is known that the conventional dipole antenna only radiates an omni-directional radiation pattern at the design frequency ω_1. As described in the previous section, the suppression of phase reversal by incorporating PMC cover has led to the improved radiation pattern (Jafargholi et al., 2010). This section is focused on the pattern modification of the wire dipole antenna using artificial magnetic conductors.

Fig. 12, shows a CLL loaded dipole antenna. The finite two-CLL-deep metamaterial Artificial Magnetic Conductor (AMC) cover was designed separately to operate at around 27GHz. The CLL dimensions were then optimized and placed optimally on both dipole arms to obtain good radiation patterns. The total number of the CLL elements is 88 (4×11×2) which symmetrically coupled to dipole antenna arms. The separation between the CLL elements is also fixed at 0.5mm.

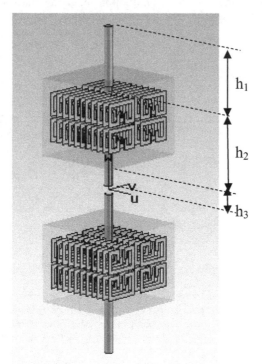

Fig. 12. A schematic view of CLL loaded dipole antenna, unit cell parameters (See Fig. 5a): L_1=0.67mm, L_2=0.89mm, L_3=1.37mm, W= 0.247mm, G=0.411mm, T_1= 0.13mm, T_2= 0.5mm, h_1= 3.25mm, h_2 =4mm, h_3 =2mm, and antenna feed gap = 0.5mm, and length of dipole antenna =19mm.

Fig. 13, shows a comparison between the reflection coefficient and input impedance of CLL loaded and unloaded wire dipole antennas of 19mm length. It was revealed in (Jafargholi et al., 2010) that ideally the dipole antenna input impedance does not change significantly by using AMC loading. However, here, the input impedance for the realized CLL loaded dipole changes due to the existence of the CLL resonance and the interaction between the CLL elements and the dipole antenna.

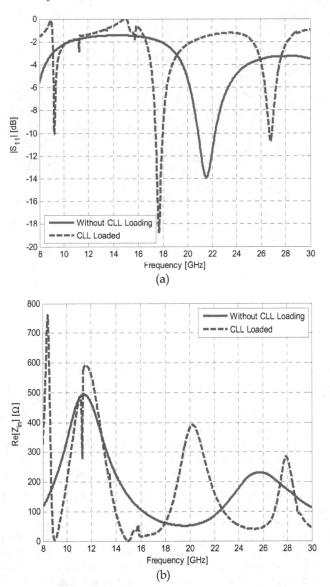

Fig. 13. (a) simulated reflection coefficient, and (b) Input Impedance of the CLL loaded and unloaded wire dipole antennas.

Fig. 14, shows the surface current densities on the CLL loaded and unloaded dipole antennas. As can be seen, the current in the CLL loaded region is the superposition of two currents oriented in opposite directions. However, the surface current caused by CLL cells is dominant, and thus the current in all parts of the dipole has the same phase. To clarify the operation mechanism of the CLL loaded dipole antenna. We assume that dipole antenna is surrounded by CLL cells, where the current direction of the dipole is reversed.

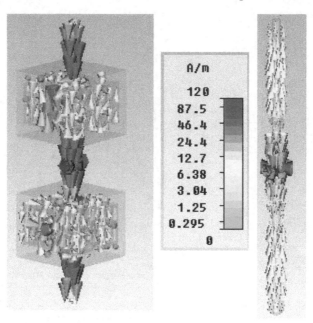

Fig. 14. Surface current densities on the CLL loaded dipole (left) and unloaded dipole (right) at *f*= 27GHz.

Fig. 15, conceptually explains the distribution of the surface current density on the CLL loaded dipole antenna. However, one must consider the complex interactions between the dipole arms and the CLL structures, such as the effects of the finite dimensions of the CLL structures on the current distribution. Consequently, full wave analysis methods have to be used in the antenna designs. However, to simplify the analysis, one can assume that the transverse dimensions of CLL loaded region are infinite in extent. Thus, as previous section, one can explain the concept based on the image theorem, i.e., when an electric current is vertical to a PMC (PEC) region, the current image has the reversed (same) direction.

For the current at the bottom of the dipole flowing into the CLL loaded region, the CLL loaded region acts as a PMC cover (Erentok et al., 2005). At the result, the direction of the image current is opposite to that of the original current, as shown in Fig. 15. In contrast, for the current at the top of dipole flowing into the CLL loaded region, the CLL loaded region acts as an AEC (Erentok et al., 2005), and thus the image current in the CLL loaded region has the same direction as the original current. The total surface current in the CLL loaded region is obtained as the sum of the two image currents and original current. At the result, the current phase of the CLL loaded dipole antenna remains unchanged through the antenna, as shown in Fig. 15.

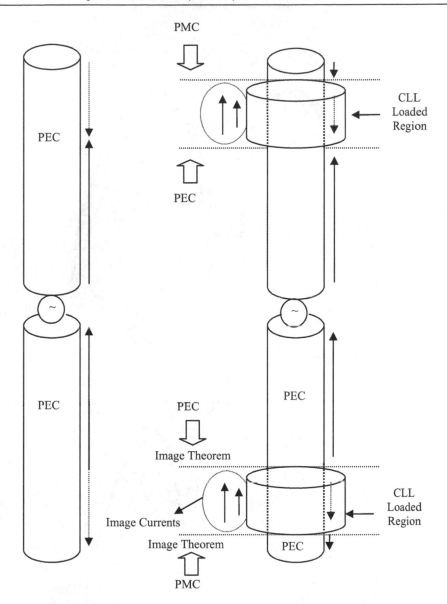

Fig. 15. Conceptual schematics of the conventional (left) and CLL loaded dipole (right).

The radiation patterns of the conventional and CLL loaded dipole antennas are also shown in Fig. 16, respectively. As can be seen, when the dipole antenna is loaded with CLL structure, the radiation patterns improve significantly, especially at the second harmonic (27GHz) of the main resonant frequency where antenna is matched well. The Simulation shows that above 20dB gain enhancement has been achieved in azimuth plane using CLL loading technique.

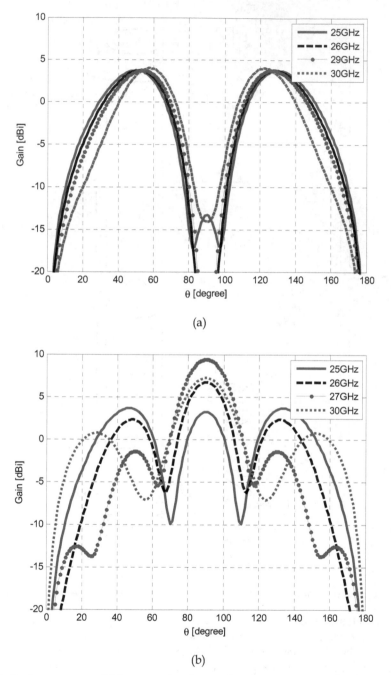

(a)

(b)

Fig. 16. Radiation patterns of (a) conventional wire dipole antenna, and (b) the CLL loaded dipole antenna versus frequency at $\varphi=0^0$.

2.4 Compact dual band loaded dipole antenna, incorporating artificial magnetic conductors

The interesting features of printed dipole antennas make it possible for them to be conformal with installation platforms such as automobiles and vessels, without affecting the aerodynamic or mechanical properties of the vehicles. In the recent years, due to unique electromagnetic properties, metamaterials have been widely considered in monopole and dipole antennas to improve their performance. However, the left-handed dipole antennas (Zhu et al., 2009; Iizuka et al., 2006; Iizuka et al., 2007), albeit compact, suffer from the losses in the loading components, namely, interdigitated capacitors and meander inductors, resulting in a very low gain.

In this section, we propose a new dual band printed dipole antenna in that CLL elements as reactive loads are placed close to the edge of the dipole. The losses associated with the CLL elements are very low in the frequency range of interest, resulting in both the acceptable gain and radiation pattern. When the CLL elements are incorporated, the resonant behavior of the unloaded printed dipole antenna changes. As a result, a new resonance is appeared with the frequency determined by the CLL dimensions. In fact, the CLL element can be easily described as an LC resonant circuit in which the resonant frequency is mainly determined by the loop inductance and the gap capacitor. It is worth noting that the resonant behavior of the CLL element starts to appear at a frequency in which the free space wavelength is much larger than its size. However, the second resonant frequency of the CLL-loaded dipole occurs at a frequency higher than the main resonant frequency of the unloaded dipole. To further reduce the second resonant frequency without increasing the area occupied by the antenna, chip capacitors are incorporated in the CLL elements. The chip capacitor provides a tuning capability of the second resonant frequency. Thus the frequency ratio between these two frequencies can be readily controlled by incorporating different chip capacitors into the capacitive gaps of the CLL elements. It is worthwhile to point out here that the CLL elements integrated with chip capacitors miniaturize the size of the printed dipole antenna.

The reason is that when the chip capacitor value is increased, the CLL resonant frequency decreases, and thus the second resonant frequency of the dipole shifts down to a lower frequency. The proposed dual band CLL loaded dipole antenna radiates effectively at both resonant frequencies with good return losses and gains as well as acceptable omnidirectional radiation patterns. The high-frequency structure simulator (Ansoft HFSS) is adopted for the simulations.

2.4.1 Dual band printed dipole antenna

In order to test the proposed approach, double-sided printed dipole antenna is loaded by CLL elements. Fig. 17, shows the CLL-loaded printed dipole antenna together with the CLL element. The dipole and CLL elements are printed on a FR4 substrate with a thickness of 0.8mm and a dielectric constant of 4.4 to reduce the cost of the antenna and to make it more rigid in construction. The CLL-loaded printed dipole has also been optimized to realize better performance. The optimized parameters of the proposed CLL-loaded dipole antenna are labeled in the Fig. 17. A prototype of the proposed CLL-loaded printed dipole is fabricated to validate the simulation results.

A photograph of the fabricated printed dipole antenna is shown in Fig. 18. Fig. 19, shows the reflection coefficient of the proposed dual band printed dipole antenna as well as the unloaded dipole antenna. As can be seen, the agreement between the simulation and measurement results is reasonably good. It is observed that when the CLL elements are added, two resonance frequencies become distinguishable from each other and thus two nulls are clearly observed in the reflection coefficient curve. The unloaded dipole antenna resonates at around 2.75GHz. In contrast, the fabricated CLL-loaded dipole resonates at 2.15 and 4.45GHz, as shown in Fig. 19. The lower resonant frequency corresponds to that of the original printed dipole and remains approximately unchanged while the higher resonant frequency is mainly due to the CLL loading. To further understand the performance of the printed dipole antenna near the CLL elements, Fig. 20, shows the magnitude of the S-parameters versus frequency for the CLL-based metamaterial.

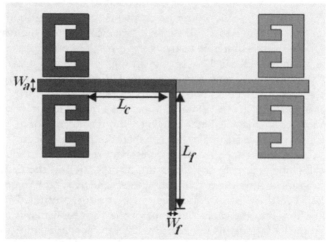

Fig. 17. Geometry of a CLL-loaded printed dipole antenna, L_f=23mm, L_c=13.67mm, W_f=1mm, W_a=2.5mm, L_1=3.73mm, L_2=4.95mm, L_3=7.62mm, G=2.28mm, W=2mm. From (Jafargholi et al., 2010), copyright © 2010 by Praise Worthy Prize, S. r. l.

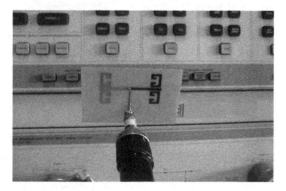

Fig. 18. Photograph of a fabricated CLL-loaded printed dipole antenna. From (Jafargholi et al., 2010), copyright © 2010 by Praise Worthy Prize, S. r. l.

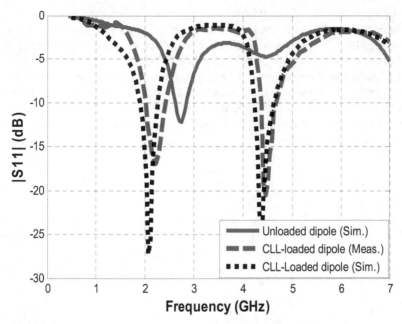

Fig. 19. Reflection coefficient comparison between the CLL-loaded and unloaded printed dipole antenna. From (Jafargholi et al., 2010), copyright © 2010 by Praise Worthy Prize, S. r. l.

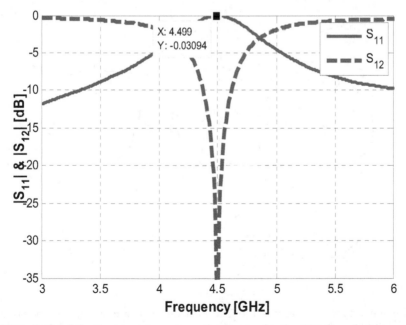

Fig. 20. Magnitude of the S-parameters versus frequency for the CLL-based metamaterial. From (Jafargholi et al., 2010), copyright © 2010 by Praise Worthy Prize, S. r. l.

It is observed that the CLL element effectively resonates at around 4.5GHz with small loss. This frequency coincides with the second resonant frequency of the CLL-loaded printed dipole (see Fig. 19). The radiation patterns of the proposed dual band printed dipole are measured at the resonant frequencies of 2.15 and 4.45GHz. Fig. 21, shows the measured and simulated E-plane radiation patterns at first and second resonant frequencies. The antenna gains at first and second resonant frequencies are 1.8dB, and 3.9dB, respectively. As a result, the losses introduced by the CLL elements are significantly low in the frequency range of interest. In other words, the proposed dual band dipole antenna has acceptable performance in both gain and radiation pattern.

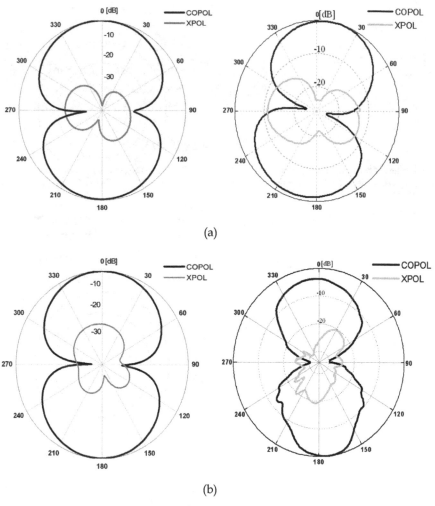

Fig. 21. E-plane radiation patterns of the CLL-loaded dipole antenna at (a) 2.15GHz, and (b) 4.45GHz. (Right hand figures are measurements). From (Jafargholi et al., 2010), copyright © 2010 by Praise Worthy Prize, S. r. l.

2.4.2 Miniaturized CLL-loaded printed dipole antenna incorporating chip capacitors

In principle, the size reduction can be arbitrarily achieved if it would be feasible to fabricate a proper metamaterial element that has a negative permeability at a frequency lower than the natural resonance frequency of the corresponding unloaded dipole antenna. For a metamaterial comprised of resonant CLL elements, this can be achieved by capacitive loading of the CLL elements. To verify and confirm the proposed approach, a prototype of a CLL-loaded dipole antenna, in which each CLL ring is loaded with a 0.68pF chip capacitor, is fabricated and measured.

A photograph of the fabricated miniaturized printed dipole antenna is shown in Fig. 22. The magnitude of the S-parameters for the CLL-based metamaterial loaded with a 0.68PF chip capacitor is shown in Fig. 23. As can be seen, the resonant frequency of the CLL element shifts down to 1.33GHz by incorporating 0.68PF chip capacitor (see Figs. 21, 23). In order to meet the specification of both the ISM system and mobile communication, the miniaturized printed dipole should radiate linearly polarized waves at 1.8GHz and 2.45GHz.

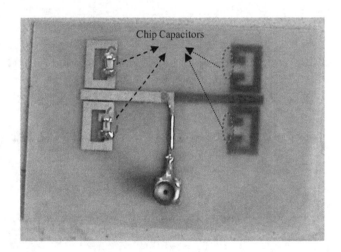

Fig. 22. Photograph of the miniaturized CLL-loaded printed dipole antenna (incorporating 0.68pF chip capacitors). From (Jafargholi et al., 2010), copyright © 2010 by Praise Worthy Prize, S. r. l.

Fig. 24, compares the measured and simulated reflection coefficient of the proposed miniaturized CLL-loaded printed dipole antenna. As can be seen, the second resonant frequency of the proposed CLL-loaded dipole of Section 1-2, considerably shifts down to a lower frequency by incorporating chip capacitors. The resonant frequencies of the proposed miniaturized CLL-loaded printed dipole are lower than the main resonant frequency of the unloaded dipole antenna. It should be pointed out that the antenna radiation patterns at both resonant frequencies are quite similar to that of a half wavelength dipole, as shown in Fig. 25. The antenna gains at first and second resonant frequencies are 0.62dB, and 2.26dB, respectively.

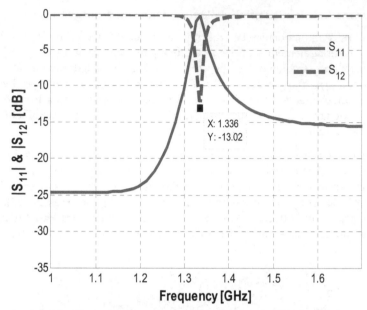

Fig. 23. Magnitude of the S-parameters versus frequency for the CLL based metamaterial loaded with a 0.68PF chip capacitor. From (Jafargholi et al., 2010), copyright © 2010 by Praise Worthy Prize, S. r. l.

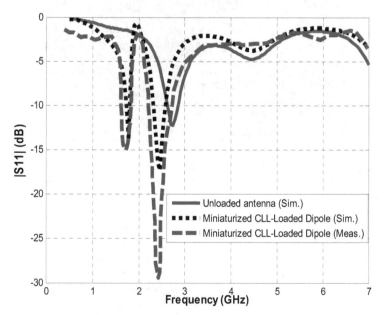

Fig. 24. Measured and simulated reflection coefficient of the miniaturized CLL-loaded printed dipole antenna, as compared to that of the unloaded printed dipole. From (Jafargholi et al., 2010), copyright © 2010 by Praise Worthy Prize, S. r. l.

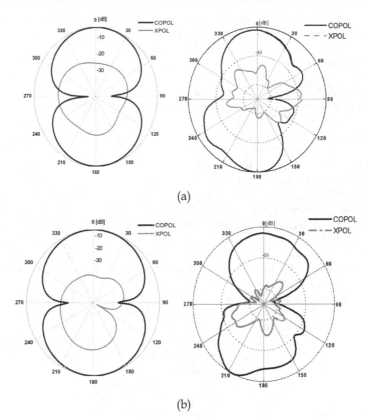

(a)

(b)

Fig. 25. E-plane radiation patterns of the miniaturized CLL-loaded printed dipole antenna of Section 1-3 at (a) 1.713GHz, and (b) 2.434GHz. (Right hand figures are measurements). From (Jafargholi et al., 2010), copyright © 2010 by Praise Worthy Prize, S. r. l.

3. Acknowledgment

The authors would like to thank Iran Telecommunication Research Centre (ITRC) for its financial supports.

4. References

[1] C. A. Balanis, *Advanced Engineering Electromagnetics*, John Wiley & Sons, New York, 1989.

[2] V. G. Veselago, "The electrodynamics of substances with simultaneously negative values of ε and μ," *Sov. Phys. Usp.*, vol. 10, no. 4, 509–514, 1968.

[3] M. Veysi, M. Kamyab, S. M. Mousavi, and A. Jafargholi "Wideband Miniaturized Polarization Dependent HIS Incorporating Metamaterials," *IEEE Antennas Wireless Propag. Letter*, vol. 9, 764-766, 2010.

[4] M. Rafaei Booket, M. Kamyab, A. Jafargholi and S. M. Mousavi, "Analytic Modeling and Implementation of The Printed Dipole Antenna loaded with CRLH Structures, " *Progressive In Electromagnetic Research B*, Vol. 20, 167-186, 2010.

[5] M. Veysi, M. Kamyab, J. Moghaddasi, and A. Jafargholi, "Transmission Phase Characterizations of Metamaterial Covers for Antenna Application," *Progressive In Electromagnetic Research Letter*, Vol.21, pp. 49-57, 2011.

[6] M. Rafaei Booket, A. Jafargholi, M. Kamyab, H. Eskandari, M. Veysi, and S. M. Mousavi, "A Compact Multi-Band Printed Dipole Antenna Loaded With Single-Cell MTM," Pending Publication in *IET Microwave Antenna Propag.*, 2011.

[7] A. Jafargholi, M. Kamyab, and M. Veysi, "PMC-based Waveguide-fed Slot Array," *ISRN Communications and Networking*, Hindawi, Vol. 2011, Article ID 941070, 5 Pages.

[8] A. Jafargholi, M. Kamyab, and M. Veysi, "Artificial Magnetic Conductor Loaded Monopole Antenna," *IEEE Antennas Wireless Propag. Letter*, vol. 9, 211-214, 2010.

[9] A. Jafargholi, M. Kamyab, M. Rafaei Booket, and M. Veysi, "A Compact Dual-band Printed Dipole Antenna Loaded with CLL-Based Metamaterials," *International Review of Electrical Engineering*, IREE, Vol. 5, No. 6, pp. 2710-2714, 2010.

[10] J. McVay, N. Engheta, and A. Hoorfar, "High-impedance metamaterial surfaces using Hilbert-curve inclusions," *IEEE Microwave Wireless Components Lett.*, vol. 14, no. 3, pp. 130–132, Mar. 2004.

[11] A. Erentok, P. Luljak, and R. W. Ziolkowski, "Antenna performance near a volumetric metamaterial realization of an artificial magnetic conductor," *IEEE Trans. Antennas Propagat.*, vol. 53, pp. 160–172, Jan. 2005.

[12] R. W. Ziolkowski and A. Kipple ,"Application of double negative metamaterials to increase the power radiated by electrically small antennas," *IEEE Trans. Antennas Propagat.*, vol. 51, no. 10, pp. 2626–2640, Oct. 2003.

[13] Q. Liu, P. S. Hall, and A. L. Borja ," Efficiency of Electrically Small Dipole Antennas Loaded With Left-Handed Transmission Lines," *IEEE Trans. Antennas Propagat.*, vol. 57, no. 10, pp. 3009–3017, Oct. 2009.

[14] S. D. Rogers, C. M. Butler, and A. Q. Martin, " Design and Realization of GA-Optimized Wire Monopole and Matching Network With 20:1 Bandwidth," *IEEE Trans. Antennas Propagat.*, vol. 51, no. 3, pp. 493–502, March. 2003.

[15] A. Erentok, and R. W. Ziolkowski, "Metamaterial-Inspired Efficient Electrically Small Antennas" *IEEE Trans. Antennas Propagat.*, vol. 56, pp. 691–707, March 2008.

[16] A. Erentok, *Metamaterial-Based Electrically Small Antennas*, Ph.D. dissertation at University of Arizona, 2007.

[17] J. Zhu, M. A. Antoniades, and G. V. Eleftheriades "A Compact Tri-Band Monopole Antenna With Single-Cell Metamaterial Loading" *IEEE Trans. Antennas Propagat.*, vol. 58, pp. 1031–1038, April. 2010.

[18] M. A. Antoniades and G. V. Eleftheriades, "A broadband dual-mode monopole antenna using NRI-TL metamaterial loading," *IEEE Antennas Wireless Propag. Lett.*, vol. 8, pp. 258–261, 2009.

[19] J. Zhu, M. A. Antoniades, and G. V. Eleftheriades, "A tri-band compact metamaterial-loaded monopole antenna for WiFi and WiMAX applications,"presented at the *IEEE Antennas and Propagation Society Int. Symp.*, Jun. 2009.

[20] J. Zhu and G. V. Eleftheriades, "Dual-band metamaterial-inspired small monopole antenna for WiFi applications," *Electron. Lett.*, vol. 45, no. 22, pp. 1104–1106, Oct. 2009.

[21] H. Iizuka, P. S. Hall, and A. L. Borja, "Dipole Antenna With Left-Handed Loading" *IEEE Antennas Wireless Propag. Lett.*, vol. 5, pp. 483–485, 2006.

[22] H. Iizuka, and P. S. Hall, "Left-Handed Dipole Antennas and Their Implementations" *IEEE Trans. Antennas Propagat.*, vol. 55, pp. 1246–1253, May 2007.

9

Compact Coplanar Waveguide Metamaterial-Inspired Lines and Its Use in Highly Selective and Tunable Bandpass Filters

Alejandro L. Borja[1], James R. Kelly[2], Angel Belenguer[1],
Joaquin Cascon[1] and Vicente E. Boria[3]
[1]*Departamento de Ingeniería Eléctrica, Electrónica, Automática y Comunicaciones,
Escuela, Politécnica de Cuenca, Universidad de Castilla-La Mancha*
[2]*School of Electronic, Electrical and Computer Engineering, University of Birmingham*
[3]*Instituto de Telecomunicaciones y Aplicaciones Multimedia,
Universidad Politécnica de Valencia*
[1,3]*Spain*
[2]*UK*

1. Introduction

During the last years, the metamaterials field has grown rapidly due to the possibility of accomplishing a methodology to achieve negative effective parameters ε_{eff} Pendry et al. (1996) and μ_{eff} Pendry et al. (1999), and their experimental verification Smith et al. (2000) - Shelby, Smith & Schultz (2001). The main research work has been concentrated on the theoretical consequences of negative parameters, as well as techniques for practically realizing left-handed media for various optical/microwave concepts and applications. Among recent concepts, properties, and devices based on engineered metamaterials, we can mention frequency selective structures, which have opened the path to a new range of passive devices for guided applications. In this context, split ring resonator (SRR) loaded transmission lines represent the cutting edge of research in the field of one-dimensional (1D) planar left-handed structures. These structures were firstly proposed by Martin et al. in 2003 Martin et al. (2003) by magnetically coupling a shunted coplanar waveguide (CPW) and pairs of SRRs. These planar devices exhibit backward propagation in a narrow frequency band above the resonant frequency of the rings, with the necessary degree of flexibility to design compact low insertion losses filters. Thus, based on this former configuration different approaches have been proposed with the aim of improving performances and overcome possible drawbacks such as asymmetrical response shape or transmission bands with smooth edges. For instance, a combined right/left-handed CPW structure was implemented by cascading SRRs-wire and SRRs-gap stages Bonache et al. (2005), satisfactorily achieving a transmission upper band with a sharp cut-off. Furthermore, new CPW lines with extra loading elements have shown interesting properties in terms of improved out-of-band behavior and response selectivity Borja, Carbonell, Boria, Cascon & Lippens (2010) - Borja, Carbonell, Boria &

Lippens (2010b). By these means, it is possible to control several restrictive trade-offs by simply adjusting the loading elements. In this way it is possible to obtain quite a symmetric frequency response along with controllable bandwidths and compact dimensions.

In this chapter, the properties of a variety of metamaterial designs are analyzed and discussed in order to develop novel small planar metamaterial frequency selective structures. The study of these configurations is based on full-wave electromagnetic analysis, equivalent circuit simulations and measured responses of different prototypes designed for microwave operation. In particular, section 2 presents different CPW lines based on a split ring resonator SRR technology and loaded with metallic strips and gaps. The properties of such structures can be controlled by properly designing or adding loading elements. In this regard, the use of shunt wires permits to control frequency selectivity by means of an engineering of the electric plasma frequency, providing deeper upper band rejection levels. On the other hand, the addition of series capacitances to previous unit cell implementations provides a transmission response which is almost symmetric while exhibiting a right-handed character along the pass band, contrary to conventional left-handed lines. In sections 3 and 4 enhanced out-of-band rejection properties and reconfigurable responses can be obtained by the use of cascaded and varactor loaded basic cells, respectively. Finally, the main conclusions of the chapter are outlined in section 5.

2. Split ring resonators based coplanar waveguide lines

In this section, different arrangements corresponding to configurations of SRRs loaded CPW lines have been considered. Two loading elements, i.e. shunt strips, and series gaps have been successively included in the CPW line. These inclusions are much smaller than the electrical wavelength of the propagating wave. Therefore, effective medium considerations in the CPW and lumped equivalent circuit models have been used to understand and better explain the complex features of wave propagation inherent to these loaded lines. In this section numerical studies and experiments are also included.

2.1 Split ring resonators loaded coplanar waveguide with shunt strips

The first model is based on the combination of SRRs and shunt wires within a host CPW line. This resonant structure has already been deeply analyzed Martin et al. (2003), and therefore it will be used as a reference result. In summary, this unit cell behaves as a left-handed propagation uni-dimensional transmission line that, if its size is sufficiently small as compared to the electrical wavelength, can have a double negative effective medium behaviour. By using the SRRs it is possible to synthesize a negative value of effective permeability, whereas shunt wires in the CPW provide a negative effective permittivity. For this reason, the structure can be considered as a double negative effective medium, when operating at frequencies slightly above the resonance of the isolated SRR Aznar et al. (2008).

The proposed left-handed structure is depicted in Fig. 1. It consists of a host CPW loaded with SRRs and two shunt strips. SRRs are symmetrically placed on the rear of the substrate, while thin metal wires connect the signal line to the ground plane at positions coincident with the center of the SRRs. Table 1 defines the CPW geometry, unit cell characteristics, and common parameter values for this and subsequent configurations.

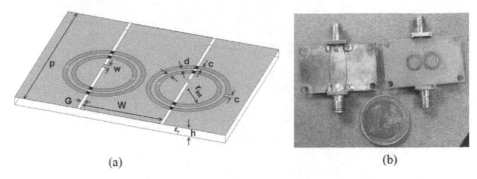

(a)   (b)

Fig. 1. (a) SRRs loaded CPW with shunt strips, and (b) photograph of the prototype.

W(mm)	G(mm)	p(mm)	w(mm)	h(mm)	ϵ_r
7.7	0.3	10	0.4	0.508	2.2

$tg\delta$	$t(\mu m)$	σ(S/m)	r_{int}(mm)	c(mm)	d(mm)
0.0009	35	$5.8 \cdot 10^7$	2.6	0.4	0.4

Table 1. Unit cell characteristics.

W is the line width, G is the gap between conductors, w is the strip width, and p the unit cell period. Substrate characteristics are height h, permittivity ϵ_r, and loss tangent $tg\ \delta$. Cooper metallization was utilized. This had a thickness t, and conductivity σ. Moreover, a prototype has been fabricated in order to verify the propagation behavior of the cell, see Fig. 1 (b). A taper section is added at both SMA connections to properly feed the device. The sample has been fabricated on a Neltec NY9220 dielectric substrate using a mechanical milling process. The milling was performed by a LPKF Protomat 93S machine. Thereafter, the fabricated device has been measured and characterized by means of a Rohde & Schwarz vector network analyzer ZVA-24, calibrated with a Through-Open-Short-Match kit, in the frequency band from 3 to 5 GHz.

In parallel, the structure proposed is analyzed and compared with the lumped element equivalent circuit of the unit cell, see Fig. 2. Due to physical symmetry properties, the magnetic wall concept has been used so that the equivalent circuit corresponds to one half of the basic cell. The equivalent circuit model can be transformed to an equivalent π-circuit type, as it was described by Aznar et al. (2008),

Each SRR can be represented by a simple LC parallel resonator circuit in the vicinity of resonance, with elements L_s and C_s. L and C are the per-section inductance and capacitance defined from the geometry of the CPW line and calculated as it is advised in Mongia et al. (1999). L_p is the equivalent inductance of connecting wires, which divides the inductances L_s and L into two parts, as proposed in Rogla et al. (2007) and thoroughly verified in Aznar et al. (2008). SRRs are modeled by parallel resonant circuits inductively coupled to the line through a coupling constant k. It is calculated by means of the fractional area theory explained in Martin et al. (2003). The values of the different lumped elements and parameters are summarized in Table 2.

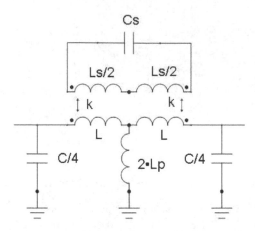

Fig. 2. Lumped equivalent circuit model of the SRRs loaded CPW with shunt strips, originally proposed in Rogla et al. (2007).

C_s (pF)	L_s (nH)	C (pF)	L (nH)	L_p (pH)	k
0.104	14.8	0.672	2.11	131	0.342

Table 2. Equivalent circuit parameters of the SRRs loaded CPW with shunt strips.

The transmission S_{21} and reflection S_{11} coefficients for the left-handed structure shown in Fig. 3 (a), have been obtained from three different sources, i. e. full-wave simulations, lumped equivalent circuit simulations, and experimental results. Agreement is found to be very good in all cases. Nonetheless, a frequency shift for the equivalent circuit response is observed. This shift could be minimized by simply tuning L_s and C_s element values, which control the SRR resonance. As it can be seen, the frequency response exhibits a pass band centred around 4.3 GHz with a transmission zero close to 3.6 GHz. According to the model of Fig. 2, the structure should exhibit a transmission zero (all injected power is returned back to the source) at that frequency where the series branch opens, and it occurs at the resonant frequency of the coupled SRRs.

The analysis of these transmission characteristics is performed by the extraction of the effective medium parameters of the uni-dimensional propagation structure, see Fig. 3 (b). The extraction is based on the well known Nicolson-Ross-Weir (NRW) procedure used in Smith et al. (2005), where the real parts of the permittivity and permeability are retrieved from the scattering parameters. The results obtained confirm the presence of a narrow pass band between 4.1 GHz and 4.5 GHz, corresponding to a double negative frequency band as it is expected. The maximum transmission is achieved when the matching condition $\epsilon \cong \mu$, and thus reduced impedance $\bar{Z} = \sqrt{\mu/\epsilon} \cong 1$, is satisfied. Henceforward, the transmission line becomes single negative as permeability reaches positive values after the magnetic plasma frequency f_{mp}. A double positive medium is subsequently obtained above the electric plasma frequency f_{ep}.

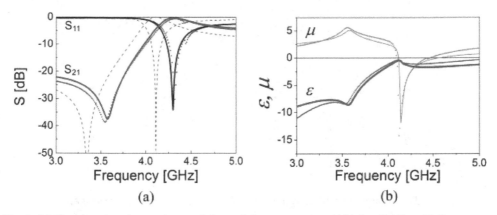

(a)   (b)

Fig. 3. (a) Simulated and experimental S_{11} and S_{21} parameters. (thick solid line: Full-wave
simulation, symbol: Measurement, thin dashed line: lumped equivalent circuit simulation).
(b) Simulated (line) and experimental (symbol) real parts of extracted permittivity (thick) and
permeability (thin) according to the NRW method.

2.2 Split ring resonators loaded coplanar waveguide with series gaps

The next model, depicted in Fig. 4 (a), is based on the combination of SRRs and series gaps.
In this configuration, the shunt wires located on the top side of the substrate have been
substituted by series gaps. These elements are dual of the shunt wires and they are located
symmetrically with regard to the center of the unit cell. The dimensions of the elements,
defined in Table 1, are the same ones as those considered in the previous configuration.
Moreover, the gap width g and separation p_1 are 0.25 mm and 5 mm, respectively. A prototype
has been manufactured in order to verify the behaviour experimentally, see Fig. 4 (b). The
sample has been fabricated using the same substrate and process described in section 2.1.

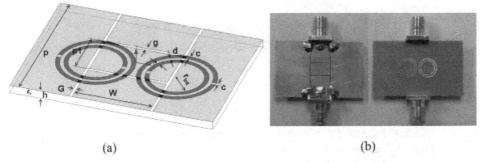

(a)   (b)

Fig. 4. (a) SRRs loaded CPW with series gaps and (b) photograph of the prototype.

Likewise, a lumped element equivalent circuit of the elemental cell, shown in Fig. 5, is used
to asses the different properties of the new configuration under study. The CPW is modeled
as described before, it includes two series capacitances which have been properly modified
accordingly to the magnetic wall theory. A gap discontinuity in a CPW line can be represented
by means of an equivalent two-port π-network, as presented by Deleniv et al. (1999). The

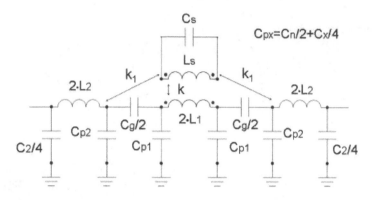

Fig. 5. Lumped equivalent circuit model of the SRRs loaded CPW with series gaps.

π-network comprises a series capacitance C_g together with a shunt capacitance C_n. The series capacitance C_g describes the reactance due to the gap discontinuity, whilst the shunt capacitance C_n accounts for grounding edge effects at the sides of the gap. L_1 and C_1 are the per-section inductance and capacitance of the line between the gaps. Also, L_2 and C_2 account for the per-section inductance and capacitance of the transmission line. C_{pi} (i= 1,2) is the equivalent capacitance of the two shunt capacitances, $C_{n/2}$ and $C_{i/4}$. The SRRs are inductively coupled to different parts of the line. Coupling to the central portion of the line, between gaps, is modeled by a coupling constant k. k is calculated applying the fractional area theory. The value of k has been adjusted slightly since the series gaps modify the coupling between the line and the rings. Following this adjustment, the bandwidth and location of the transmission zero can easily be tuned. Similarly, the coupling constant k_1 represents the interaction between the external portions of the CPW and the SRRs. The value of k_1 was adjusted by means of a curve fitting procedure, which ensured that the equivalent circuit simulations agreed well with the measurement results. Table 3 gives the final values of the circuit elements within the equivalent circuit.

C_s (pF)	C_g (pF)	C_n (pF)	C_1 (pF)	C_2 (pF)
0.104	0.3	0.0485	0.3192	0.1596

L_s (nH)	L_1 (nH)	L_2 (nH)	k	k_1
14.8	1	0.5	0.6	0.35

Table 3. Equivalent circuit parameters of the SRRs loaded CPW with series gaps.

Fig. 6 (a) shows the scattering parameters obtained through simulation and measurement. A good agreement between the three responses, i.e. full-wave, equivalent circuit, and measurements is observed. As a relevant feature, it can be mentioned that the structure exhibits a transmission band centred around 3.8 GHz and a transmission zero around 4.5 GHz. This is contrary to the performance of a CPW line loaded with SRRs and shunt strips, where the transmission zero is located below the pass band, see Fig. 3 (a). In this structure the transmission zero is located at high frequencies and the pass band appears before the anti-resonance. An important reminder here is that the SRRs are exactly the same in these two configurations, and only the other loading elements (series gaps and shunt wires, not resonant

by themselves in this frequency range) are different. The parallel inductive contribution of the
wires is replaced by a series capacitance. For this reason the device exhibits dual behaviour
compared to a CPW line loaded with SRRs and shunt strips.

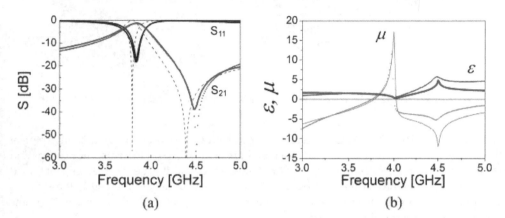

(a)   (b)

Fig. 6. (a) Simulated and experimental S_{11} and S_{21} parameters. (thick solid line: Full-wave
simulation, symbol: Measurement, thin dashed line: lumped equivalent circuit simulation).
(b) Simulated (line) and experimental (symbol) real parts of extracted permittivity (thick) and
permeability (thin) according to the NRW method.

This performance can be explained by the interpretation of the retrieved permittivity and
permeability of the structure, presented in Fig. 6 (b). The generation of a pass band is related to
a double-positive condition. Permittivity remains positive in the whole measured frequency
band. Permeability is also positive in a small frequency range, below the resonant frequency
of the SRR (4 GHz), and negative with a monotonous variation outside it. In this case, the
SRR is contributing to generate a right-handed transmission band. This is due to the series
capacitance loaded in the line, which precludes transmission outside of the area where the SRR
resonates. Permittivity is also affected by the presence of the SRR, but it remains positive as
mentioned before. Note also, as expected, that peak transmissions correspond to the different
crossings of the ϵ and μ curves ($\epsilon \cong \mu$ gives the matching condition and hence maximum
transmission and minimum reflection). The transmission zero in Fig. 6 (b) is located coincident
with the positions of the slope changes in the effective parameters. These slope changes are in
turn due to the superposition of the 'monotonous' behaviours of the simple loading elements
(series gaps) with the complex (quasi-Lorentz-type) behaviour of the SRR loaded in the line.

2.3 Split ring resonators loaded coplanar waveguide with shunt strips and series gaps

At last, the characteristics of transmission lines combining all the previous elements, shunt
strips and series gaps in CPW technology are studied. The unit cell, described in Fig. 7
(a), consists of an SRRs based CPW loaded with two shunt metallic strips and two series
capacitances.

The gaps are symmetrically placed with respect to the SRRs, while the thin shunt strips
connecting the central line to the ground are placed at those positions coincident with the

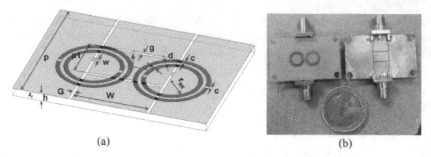

(a)   (b)

Fig. 7. (a) SRRs loaded CPW with shunt strips and series gaps, and (b) photograph of the prototype.

center of the SRRs. The series gap elements can be interpreted as the dual counterpart of the shunted inductive strips as it was discussed previously. These elements, shunt strips and series gaps, are in the coupling regions of the SRRs so that, as it is next explained, combined effects of previous properties take place. The configuration of the unit cell is the same as that employed in previous sections. The dimensions are given in Table 1. In Fig. 7 (b), the prototype of the unit cell is depicted.

The lumped equivalent circuit model of the basic cell, which will be used for the interpretation of the structure, is presented in Fig. 8. As it was reported previously, due to symmetry, the magnetic wall theory has been applied. In addition, the different elements of the circuit model are identical to those described in sections 2.1 and 2.2 (see also Table 4).

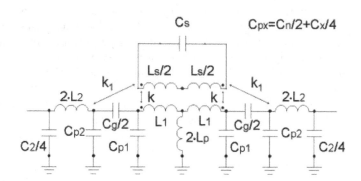

Fig. 8. Lumped equivalent circuit model of the SRRs loaded CPW with shunt strips and series gaps.

C_s (pF)	C_g (pF)	C_n (pF)	C_1 (pF)	C_2 (pF)
0.104	0.44	0.0485	0.588	1.3

L_s (nH)	L_1 (nH)	L_2 (nH)	L_p (pH)	k	k_1
14.8	1.84	4.08	131	0.6	0.1

Table 4. Equivalent circuit parameters of the SRRs loaded CPW with shunt strips and series gaps.

The simulated and measured frequency responses for the proposed structure are shown in Fig. 9 (a). An excellent agreement between full-wave simulation data and experimental results can be observed. There is also satisfactory agreement between the results obtained using the proposed equivalent circuit and those derived by other means.

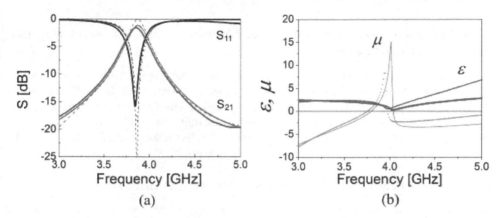

Fig. 9. (a) Simulated and experimental S_{11} and S_{21} parameters. (thick solid line: Full-wave simulation, symbol: Measurement, thin dashed line: lumped equivalent circuit simulation). (b) Simulated (line) and experimental (symbol) real parts of extracted permittivity (thick) and permeability (thin) according to the NRW method.

In the two cases previously studied, namely SRR loaded CPW with shunt strips and SRR loaded CPW with series gaps, a highly asymmetric frequency response with a characteristic anti-resonance effect (dip in the transmission) was obtained. In contrast, the frequency dependence of the cell with strips and gaps simultaneously exhibits an almost symmetrical pass band response centred around 3.9 GHz, just below the intrinsic SRR resonance. There are no transmission zeros in the vicinity of this pass band.

The third structure exhibits better selectivity than the previous ones, where it was only improved at frequencies above the pass band. The absence of transmission zeros is attributed to the transmission levels of each single element. In a CPW line loaded by SRRs, the shunt strips generate a transmission zero at lower frequencies but permit transmission above pass band. For this reason, the transmission zero introduced by the gaps disappears. In the same way, the transmission zero introduced by the strips is cancelled by the high transmission levels due to the series gaps.

The pass band is associated with the double positive condition (positive permittivity and permeability), see Fig. 9 (b). This condition is only achieved over a very narrow range of frequencies around, 3.9 GHz. The absence of transmission zeros can be attributed to the lack of slope changes in the effective parameters. For the models studied in sections 2.1 and 2.2, the transmission zeros in Fig. 3 and Fig. 6 are located coincident with the position of the slope changes in the effective parameters. These slope changes are due to a superposition of effects caused by the simple loading elements (series gaps and shunt wires) and the SRR. In common with the CPW line loaded with SRRs and gaps, this structure has a right-handed behaviour.

For this reason one can conclude that the gaps effect would dominate over that of the shunt strips. Whenever shunt strips are used there will be a transmission zero in the lower part of the frequency spectrum. This is the only effect using shunt strips. The frequency response will be symmetric and right-handed if the gaps are present. Additionally, as it can be expected, peak transmission corresponds to the matched condition $\epsilon \cong \mu$.

3. Compact and highly selective left-handed transmissions lines loaded with split ring resonators and wide strips

In the present section, it will be shown numerically and experimentally that problems related to out-of-band rejection can be alleviated by a proper arrangement of the loading elements responsible of the electrical response (shunt wires). Also, it is demonstrated that the selectivity of the transmission window can be improved by cascading basic cells. This opens up the possibility to fabricate band pass filters based on the SRR technology with excellent trade-offs between selectivity, insertion losses, and out-of-band rejection.

3.1 Split ring resonators loaded coplanar waveguide with wide shunt strips

The electromagnetic properties of left-handed materials, which are highly dispersive due to the transition between the single negative and double negative conditions, have shown interesting frequency filtering properties. These properties rely on the same physical principle, namely the magnetic coupling of a transmission medium to micro-resonators. This effect is responsible for producing a negative effective permeability above the resonant frequency of the resonators. On the other hand, an arrangement of shunt strips create a medium which exhibits negative values of the effective permittivity (ϵ_{eff}) and a high pass filter response. The overlap of the spectrum where ϵ_{eff} and μ_{eff} are simultaneously negative gives a frequency band in which the wave propagation is backward. This correspond to the so-called Left-Handed (LH) rule in terms of E, H and k trihedron. Indeed, it is now well established that the dispersion properties of the negative effective permittivity and permeability are very different, with a Drude-like and Lorentz-type frequency variation, respectively (see Fig. 10).

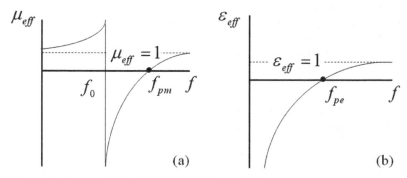

Fig. 10. (a) Lorentz-type model and (b) Drude-like model.

In short, whereas the permittivity increases continuously from negative values to positive
ones, at the crossing point known as the electrical plasma frequency, the variations of the
effective permeability versus frequency show a resonant response. The two frequencies
involved are the resonance frequency of resonators and the magnetic plasma frequency. As a
consequence, the asymmetric and double negative overall response is the superposition of a
resonant transmission onto a baseline that increases with frequency. Under these conditions,
the electrical plasma frequency which defines the transition between the negative and positive
value of ϵ_{eff} is generally adjusted until it is slightly higher than the magnetic plasma
frequency. By this choice it is generally believed that the impedance matching conditions
can be met with comparable values of μ and ϵ and hence impedance $Z \approx 1$.

An alternative technique for improving the selectivity of the transmission window is proposed
below, without any additional loading elements. It is based on the engineering of the electric
plasma frequency f_{pe}. The idea is to increase f_{pe} with respect to the magnetic plasma
frequency f_{pm}. In this way it is possible to ensure that the material is single negative over
a broader range of frequencies. Consequently, the rejection level is increased in the upper
part of the spectrum. In addition, the impedance matching conditions are not significantly
degraded due to the fact that the effective permittivity is also influenced by the resonance
effect of the SRRs. For this reason, the values of ϵ_{eff} are still quite comparable to those of
μ_{eff} in the vicinity of the resonance. F_{pe} is tailored by enlarging the width of the shunt strip.
This also has the effect of dramatically decreasing the coupling between the resonator and the
CPW line. As a consequence, the selectivity of the left-handed pass band is enhanced. The
model proposed in section 2.1 is further analyzed in order to improve its frequency response.
Fig. 11 (a) shows a general schematic of the unit cell.

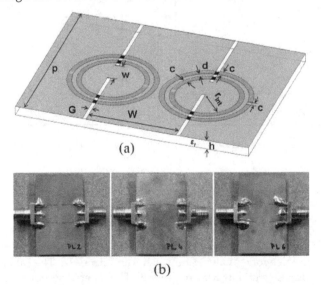

(a)

(b)

Fig. 11. (a) SRRs loaded CPW with wide shunt strips configuration, and (b) photograph of
prototypes with widths w=2, 4 and 6 mm, respectively.

The dimensions and characteristics are given in Table 1. Additionally, Fig. 11 (b) shows three fabricated prototypes with identical dimensions, only w has been modified for the different devices. The width of connecting wires is varied from w=2 mm to w=6 mm, in 2 mm steps. In previous implementations, w was approximately equal to the SRR strip width c (where c=0.4 mm). In this case, w was increased by a factor of 10. With these SRR dimensions, the unloaded Q factor calculated using the eigenmode solver in HFSS, was found around 400. This assumes a copper conductivity of $\sigma = 5.8 \cdot 10^7$ S/m and dielectric loss tangent of tg δ=0.0009. Next, the simulated and measured responses are compared. Fig. 12 shows the scattering parameters S_{11} and S_{21} as a function of frequency for the four single-cell prototypes, which simply differ by the width of the shunt strip w as aforementioned.

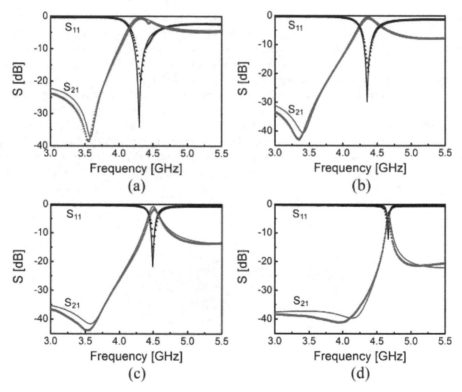

Fig. 12. Comparison between measured (symbol) and calculated (solid lines) scattering parameters for (a) w=0.4 mm, (b) w=2 mm, (c) w=4 mm and (d) w=6 mm.

In Fig. 12 (a) the width was set to w=0.4 mm, while in Fig. 12 (b), (c), and (d) it was increased to w=2 mm, w=4 mm, and w=6 mm respectively. In practice, the S_{ij} parameters were measured by means of a Vector Network Analyzer between 3 and 5.5 GHz. The calibration procedure is based on a Through-Open-Short-Match method. The tapered sections which interconnect the coaxial connector and the basic cell have been removed by a de-embedding process. Comparing Fig. 12 (a), (b), (c), and (d), it can be seen that enlarging the strip width by one order of magnitude, keeping the SRR geometry unchanged, has different major consequences:

an increase of the selectivity, a decrease of the bandwidth, a shift of the resonant frequency, and an increase of the insertion losses.

First of all, it can be noticed that the out-of-band rejection maximum at around 5 GHz which corresponds to the transmission level where the derivative of S_{21} is vanishing, shows a huge increase. Rejection levels are shown in Table 5. Let us keep in mind that such a high rejection level was obtained with a single cell. The measured return losses are equal to -20 dB for w=0.4 mm at around 4.25 GHz, while they reach -15 dB for w=4 mm at around 4.5 GHz, showing that the mismatch degradation is moderate, and compatible with real life applications where return loss around -10 dB can be enough.

	w=0.4 mm	w=2 mm	w=4 mm	w=6 mm
Out-of-band rejection (dB) at 5 GHz	−4.5	−7.6	−13	−22.7
FBW (%)	12.7	7.2	3.3	1.2

Table 5. Shunt strip effect on the out-of-band rejection and FBW.

Furthermore, it can be shown that the bandwidth at half maximum of the transmission is considerably narrower with a fractional bandwidth (FBW) decreasing from $FBW = 12.7\%$ for w=0.4 mm to $FBW = 1.2\%$ for w=6 mm, see Table 5.

Despite the fact that the SRR dimensions were kept unchanged between the four prototypes, a slight shift in the resonance frequency was observed. It is not due to fabrication tolerances, owing to the good fit with the full-wave simulation results. This shift is attributed to a slight alteration of the SRRs excitation, as a consequence of the modification of the strip. The interaction between SRRs and shunt strips is what determines in practice the position of the maximum transmission peak in the left-handed transmission band. Thus, as long as the width w is varied, the transmission band is shifted. Finally, it can be noted that there is a moderate increase in the insertion losses due to the increase in width of the shunt strip. It is for the reason that very selective frequency responses are deeply affected by the conductor and dielectric losses. Also, SRRs are weakly excited.

In order to have further insight into the electromagnetic properties in terms of dispersion characteristics, the frequency dependence of the effective parameters has been retrieved. To this aim, the Nicolson-Ross-Weir procedure, which has also been applied in section 2, is used. It is worthwhile to mention that the retrieval process is usually utilized when the dimensions of the unit cell are electrically small compared to the wavelength, generally $\lambda_g/10$. It means that when the working frequency increases the extracted parameters are less accurate. The applicability of such methods are strongly related to the frequency range where the device is considered to be used. Therefore, the results extracted from this analysis method should be considered as an estimation rather than an exhaustive study.

Fig. 13 shows the frequency dependence of the real parts of the effective permittivity and permeability for w=0.4 mm and w=4 mm. For simplicity, two values are shown, being enough to verify the effect of the shunt strip width increase. The key result is the observation of a shift of the frequency when ϵ_{eff} goes from the negative values to positive ones, this corner frequency corresponding to f_{pe}. Quantitatively it can be seen that f_{pe} is shifted from 6.6 GHz to 8.8 GHz for simulations, and from 5.2 GHz to 7.4 GHz for measurements, when the strip width is widened by an order of magnitude starting from w=0.4 mm. It is also important to note that a resonant feature is superimposed on the conventional Drude-like variation ($\epsilon = 1 - (\omega_{pe}/\omega)^2$) envelope, helping to keep good input impedances. ϵ_{eff} reaches values comparable to the values of μ_{eff} due to the resonant effect of the SRRs, thus having in both cases good impedance matching levels. On the contrary, the dispersion of the effective permeability is less affected by the increase of the strip width.

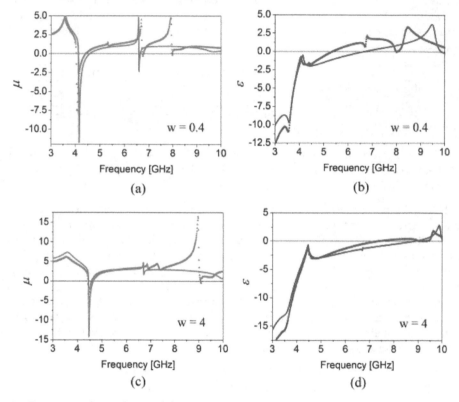

(a)   (b)

(c)   (d)

Fig. 13. Frequency dependence of the real parts of the permeability ((a),(c)) and effective permittivity ((b),(d)) for w=0.4 mm and w=4 mm, respectively. Simulation (line) and experimental (symbol) results.

The characteristic frequencies of a Lorentz dispersion law, f_o and f_{pm} respectively, remain practically unchanged by the increase of w. However, a slight frequency shift and narrower window with negative permeability values is observed, which is a consequence of the modified interaction between the SRRs and the wide shunt strip. Derived from these results, wide shunt strips present higher electric plasma frequencies, and consequently deeper rejections in the out-of-band are achieved. More to the point, through this shunt strip enlarging, the coupling between the CPW and the SRRs is reduced, providing narrower bandwidth and higher insertion losses. Both effects are interdependent, and the alteration of w brings always together a modified rejection in the out-of-band and coupling levels between the SRRs and the line. In both cases the comparison of the measured and calculated

data, shows a relatively good agreement. However, some discrepancy can be noticed. This
discrepancy is attributed to different reasons. Two taper sections are used in the fabricated
devices, so in order to retrieve effective parameters they have to be de-embeeded. This is a
possible source of errors. At the same time, finite size ground planes, fabrication tolerances,
and the measurement procedure also introduce variations in the experimental response.

3.2 Array of split ring resonators loaded coplanar waveguide with wide shunt strips

The previous section has shown that there is a strong relationship between rejection in the
out-of-band region, bandwidth, insertion losses, and selectivity as a function of the shunt
strip width. Devices with a high selectivity and narrow bandwidth (high Q) generally present
high values of insertion losses. In contrast, devices with moderate or low insertion losses do
not have sufficiently selective responses to accomplish design requirements with restrictive
out-of-band rejection or narrow bandwidth. Therefore, the key aspect in narrow band pass
filters is the necessity of a trade-off between achievable insertion losses in the pass band and
required frequency selectivity. The advantage of using the proposed SRRs loaded CPW line
with wide strips is the possibility to fulfill this trade-off, as will be shown, when several unit
cells are connected. The topology of the structure and a fabricated prototype are presented in
Fig. 14.

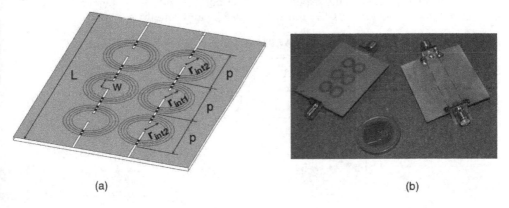

(a)   (b)

Fig. 14. Model of the SRRs loaded CPW line with wide strips composed of 3 cascaded unit
cells.

The proposed structure consists of three optimized stages, where the internal radius r_{int1} and
r_{int2}, and the unit cell period p have been modified in order to achieve a reflection coefficient
$S_{11} < -10$ dB along the pass band for two different strip widths w=3.5 mm and w= 4.5
mm. After optimization, parameter values used in the process have been set to r_{int1}=2.6
mm, r_{int2}= 2.58 mm, and p=10 mm for w=3.5 mm, while r_{int1}=2.61 mm, r_{int2}=2.59 mm, and
p=9.9 mm for w= 4.5 mm. The total length of the device is L=30 mm. The rest of the cell
parameters are the same as those given in Table 1. In Fig. 15 the frequency response of the
two optimized filters is shown. In addition, Table 6 summarizes the main properties of the
proposed filters. It can be seen that the response of both structures shows excellent insertion
loss levels for the two different widths used even if narrow pass bands are achieved. Besides,

good rejections levels, better than -30 dB, below and above the pass band are obtained with small ripple characteristics in the pass band of 0.4 dB and 0.6 dB. The main advantages of the filters proposed are the small insertion losses obtained in very selective filters with compact dimensions. In particular, the advantage of miniaturization when SRRs loaded CPW are implemented can be clearly appreciated in Fig. 16.

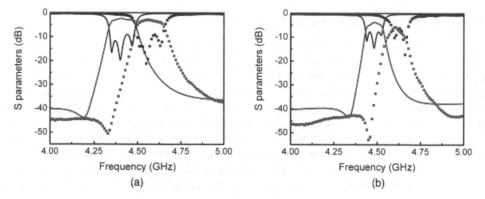

Fig. 15. Simulated (solid line) and experimental (symbol) S_{11} and S_{21} parameters for the optimized 3 stage SRRs loaded CPW line with (a) w=3.5 mm, and (b)w=4.5 mm.

	FBW (%)	IL (dB)	Ripple (dB)
w=3.5 mm	3.9	1.3	0.4
w=4.5 mm	1.8	2.2	0.6

Table 6. Characteristics of the optimized 3 stage SRRs loaded CPW simulated response.

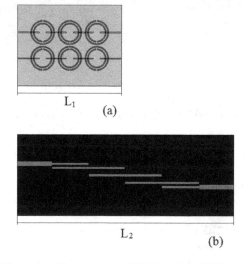

Fig. 16. Comparison of the order-3 layouts. (a) SRRs loaded CPW with wide shunt wires filter, and (b) edge-coupled line filter.

It compares the dimensions of two simulated models, namely the SRRs loaded CPW with wide shunt strips and a conventional edge-coupled lines filter. The total length L_1 is roughly 30 mm (three times the unit cell period), whereas the length L_2 is approximately 68 mm. The proposed filters are shortened by a factor of 2.3. The length of the edge-coupled filter could be further reduced by bending the coupled half-wavelength resonators in a U-shape. However, the order, and thus the length, of the filter cannot be reduced to keep good out-of-band rejection levels. On the contrary, the size of the proposed filter can still be reduced by enlarging the width of the shunt strips. Also, a comparison in terms of S-parameters is shown in Fig. 17. As it can be observed, both filters have a similar behavior. Attenuation levels outside pass band are comparable up to -30 dB for both cases. Also, a slight increment of approximately 1.1 dB in the insertion loss level can be observed for the edge-coupled line filter.

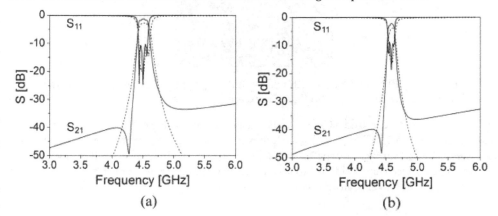

Fig. 17. Simulated S_{11} and S_{21} parameters for the optimized 3 stage SRRs loaded CPW line with (a) w=3.5 mm, and (b)w=4.5 mm. Also, the simulated S-parameters for a conventional order-3 edge-coupled filter with similar performance are depicted (dashed line).

4. Synthesis of compact and highly selective filters with tunable responses

This section presents a tunable filter based on the metamaterial transmission lines incorporating dispersive cells. Generally, in order to generate reconfigurable devices, individual reconfigurable components such as tunable capacitors and resonators are used. In particular, this reconfigurable capability is generated by an alteration of the SRRs' resonant frequency using reverse-biased varactors. Subsequently, the adjustable resonant frequency rings are employed to load the host transmission lines.

4.1 Tunable basic cell configuration

Fig. 18 shows the sketch of the varactor diode loaded SRR implemented in a CPW based configuration. The varactor is a reverse-biased semiconductor diode connected between the concentric rings of the SRR. Its capacitance can be tuned by changing the DC voltage applied to its pads. Fig. 19 shows the simulated scattering responses of the basic cell for different values of applied bias voltages (different capacitances and thus SRR resonant frequencies). The tuning range obtained is roughly 1.1 GHz with insertion losses of 2.7 dB for the higher

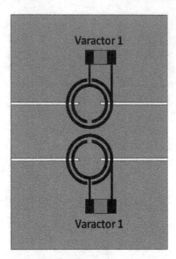

Fig. 18. Basic cell CPW line loaded with wide shunt strips and SRRs based diode varactors. Filter dimensions are the same as in figure 1, except w = 4 mm.

pass band response at around 5 GHz (f_{max}), and a maximum value of 3.8 dB when the tunable pass band is shift to 3.9 GHz ($0.78 f_{max}$). Moreover, as observed, rejection above the pass band increases as the operating frequency decreases. Compared to other tunable cell configurations, this approach effectively yields a broader frequency tuning range whilst preserving good insertion losses and a very narrow bandpass response.

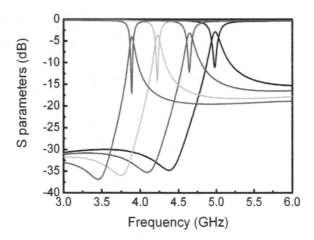

Fig. 19. Simulated scattering parameters of the tunable basic cell.

4.2 Three-cells array of split ring resonators loaded coplanar waveguide with varactors

Next, a synthesis technique for coupled resonators is applied to design a reconfigurable filter. We used the generic coupled resonators scheme of an N-order band pass filter structure by

simply cascading the previous tunable basic cell. A third order reconfigurable band pass filter
using varactor loaded SRRs, with a pass band centred between 4 and 5 GHz is considered.
The layout of the filter is shown in Fig. 20.

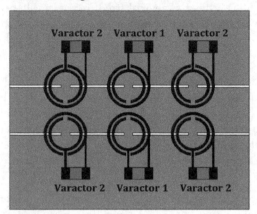

Fig. 20. Three-stage tunable filter loaded with wide shunt strips and diode varactors. Filter
dimensions are the same ones as in figure 1, except w=2mm.

The design is performed by optimizing independently the central and side stages. In this
regard, different loaded SRRs cells are used, with independent diode varactors loading
elements as depicted in Fig. 20. The simulated S-parameter results of the third order tunable
bandpass filter are shown in Fig. 21. The simulated filter can be tuned from 3.9 GHz to 4.9 GHz
(approximately 25% variation), by changing the biasing voltage from 0 to 25 V. The simulated
insertion losses at 3.9 GHz are 4.3 dB and 3.2 dB at 4.9 GHz. For the applications where
such difference is acceptable, the bandwidth of the filter is calculated to be 2.6% and 5%,
respectively.

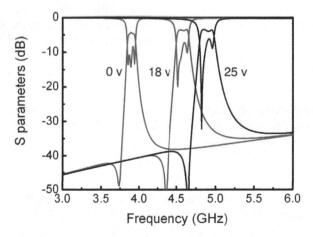

Fig. 21. Optimized simulated response of the 3-stage tunable filter.

Despite the slight increment of the insertion losses in the lower frequency band, the tuning range of the varactor loaded SRR filter is fairly good, and its size is roughly 2.5 times smaller than conventional edge-coupled filters working in the same frequency range. In addition, using a single tunable filter instead of several fixed-frequency filter bands can add system flexibility, which may warrant this slightly augmented insertion losses.

5. Summary

In this chapter, different types of coplanar transmission lines loaded with SRRs, shunt wires, series gaps, diode varactors and a combination of them have been analyzed. When different loading elements are added, the left-handed behavior originally predicted for these microstructures can be modified, and their frequency selectivity can be enhanced while maintaining the advantage of miniaturization. The analysis has been performed through full-wave electromagnetic simulations, lumped equivalent circuit models, and measurements of different prototypes.

In section 2, the propagation features of loaded CPW lines have been presented. Left-handed or right-handed propagation can be achieved depending on the loading elements included in the line, using SRRs as a common element. The use of series gaps, and their combination with shunt wires, increases the possibilities of generating narrow transmission bands in the vicinity of the SRR resonance. For instance, they can be tailored to even obtain symmetrical frequency responses. In conclusion, the proposed structures offer an alternative for designing planar frequency filtering structures in applications with severe restrictions in terms of rejection, selectivity, and size.

Section 3 has shown that it is possible to improve the selectivity of highly dispersive transmission lines, made of CPWs loaded with SRRs and shunt wires. This can be achieved by a proper engineering of the electric plasma frequency with respect to the magnetic plasma frequency. Beyond a higher rejection level, which was expected due to the deepening of the forbidden gap between the left- and right-handed dispersion branches, there is also a huge enhancement of the loaded Q quality factors. The structure also maintains low insertion losses. By cascading elementary cells it is possible to achieve a further increase of the steepness of the rejection with low insertion losses. Finally, section 4 has demonstrated reconfigurable filters having narrow bandpass responses, good insertion loss, and good frequency tuning range. The devices were based on varactor diode loaded SRRs and rigorous optimization processes.

Potential use of these miniaturized high-Q frequency selective cells can be foreseen in many modern microwave areas, notably in automotive, radar, wireless communication systems and biosensors.

6. References

Aznar, F., Bonache, J. & Martin, F. (2008). Improved circuit model for left-handed lines loaded with split ring resonators, *Applied Physics Letters* 92(4): 043512.

Bonache, J., Martin, F., Falcone, F., Garcia, J., Gil, I., Lopetegi T.and Laso, M. A. G., Marques,
R., Medina, F. & Sorolla, M. (2005). Compact coplanar waveguide band-pass filter at
the S-band, *Microwave and Optical Technology Letters* 46(1): 33–35.

Borja, A. L., Carbonell, J., Boria, V. E., Cascon, J. & Lippens, D. (2010). Synthesis of
compact and highly selective filters via metamaterial-inspired coplanar waveguide
line technologies, *IET Microwaves, Antennas and Propagation* 4(8): 1098–1104.

Borja, A. L., Carbonell, J., Boria, V. E. & Lippens, D. (2008). Synthesis of compact and highly
selective filters via metamaterial-inspired coplanar waveguide line technologies,
Applied Physics Letters 93: 203505.

Borja, A. L., Carbonell, J., Boria, V. E. & Lippens, D. (2010a). A 2% bandwidth C-band filter
using cascaded split ring resonators, *IEEE Antennas and Wireless Propagation Letters*
9: 256–259.

Borja, A. L., Carbonell, J., Boria, V. E. & Lippens, D. (2010b). A compact coplanar waveguide
metamaterial-inspired line and its use in tunable narrow bandpass filters, *40th
European Microwave Conference (EuMC), 26 Sep. - 1 Oct., Paris*.

Borja, A. L., Carbonell, J., Boria, V. E. & Lippens, D. (2009). Highly selective left-handed
transmission line loaded with split ring resonators and wires, *Applied Physics Letters*
94: 143503.

Carbonell, J., Borja, A. L., Boria, V. E. & Lippens, D. (2009). Duality and superposition in split
ring resonator loaded planar transmission, *IEEE Antennas and Wireless Propagation
Letters* 8: 886–889.

Deleniv, A., Vendik, I. & Gevorgian, S. (1999). Modeling gap discontinuity in coplanar
waveguide using quasistatic spectral domain method, *2000 John Wiley & Sons,
International Journal on RF and Microwave* 10: 150–158.

Martin, F., Bonache, J., Falcone, F., Sorolla, M. & Marques, R. (2003). Split ring resonator-based
left-handed coplanar waveguide, *Applied Physics Letters* 83(22): 4652–4654.

Mongia, R., Bahl, I. & Bhartia, P. (1999). RF and microwave coupled-line circuits, *Artech House,
Boston* .

Pendry, J. B., Holden, A. J., Robbins, D. J. & Stewart, W. J. (1999). Magnetism from conductors
and enhanced nonlinear phenomena, *IEEE Transactions on Microwave Theory and
Techniques* 47(11): 2075–2084.

Pendry, J. B., Holden, A. J., Stewart, W. J. & Youngs, I. (1996). Extremely low frequency
plasmons in metallic mesostructures, *Physical Review Letters* 76: 4773–4776.

Rogla, L. J., Carbonell, J. & Boria, V. E. (2007). Study of equivalent circuits for open-ring and
split-ring resonators in coplanar waveguide technology, *IET Microwaves, Antennas
and Propagation* 1(1): 170–176.

Shelby, A., Smith, D. R. & Schultz, S. (2001). Experimental verification of a negative index of
refraction, *Science* 292: 7779.

Shelby, R. A., Smith, D. R., Nemat-Nasser, S. C. & Schultz, S. (2001). Microwave transmission
through a two-dimensional, isotropic, left-handed metamaterial, *Applied Physics
Letters* 78(4): 489–491.

Smith, D. R. & Kroll, N. (2000). Negative refractive index in left-handed materials, *Physical
Review Letters* 85: 2933–2936.

Smith, D. R., Padilla, W. J., Vier, D. C., Nemat-Nasser, S. C. & Schultz, S. (2000). Composite medium with simultaneously negative permeability and permittivity, *Physical Review Letters* 84(18): 4184–4187.

Smith, D. R., Vier, D. C., Koschny, T. & Soukoulis, C. M. (2005). Electromagnetic parameter retrieval from inhomogeneous metamaterials, *Physical Review E* 71(1): 036617.

Permissions

The contributors of this book come from diverse backgrounds, making this book a truly international effort. This book will bring forth new frontiers with its revolutionizing research information and detailed analysis of the nascent developments around the world.

We would like to thank Xun-Ya Jiang, for lending his expertise to make the book truly unique. He has played a crucial role in the development of this book. Without his invaluable contribution this book wouldn't have been possible. He has made vital efforts to compile up to date information on the varied aspects of this subject to make this book a valuable addition to the collection of many professionals and students.

This book was conceptualized with the vision of imparting up-to-date information and advanced data in this field. To ensure the same, a matchless editorial board was set up. Every individual on the board went through rigorous rounds of assessment to prove their worth. After which they invested a large part of their time researching and compiling the most relevant data for our readers. Conferences and sessions were held from time to time between the editorial board and the contributing authors to present the data in the most comprehensible form. The editorial team has worked tirelessly to provide valuable and valid information to help people across the globe.

Every chapter published in this book has been scrutinized by our experts. Their significance has been extensively debated. The topics covered herein carry significant findings which will fuel the growth of the discipline. They may even be implemented as practical applications or may be referred to as a beginning point for another development. Chapters in this book were first published by InTech; hereby published with permission under the Creative Commons Attribution License or equivalent.

The editorial board has been involved in producing this book since its inception. They have spent rigorous hours researching and exploring the diverse topics which have resulted in the successful publishing of this book. They have passed on their knowledge of decades through this book. To expedite this challenging task, the publisher supported the team at every step. A small team of assistant editors was also appointed to further simplify the editing procedure and attain best results for the readers.

Our editorial team has been hand-picked from every corner of the world. Their multi-ethnicity adds dynamic inputs to the discussions which result in innovative outcomes. These outcomes are then further discussed with the researchers and contributors who give their valuable feedback and opinion regarding the same. The feedback is then collaborated with the researches and they are edited in a comprehensive manner to aid the understanding of the subject.

Apart from the editorial board, the designing team has also invested a significant amount of their time in understanding the subject and creating the most relevant covers. They scrutinized every image to scout for the most suitable representation of the subject and create an appropriate cover for the book.

The publishing team has been involved in this book since its early stages. They were actively engaged in every process, be it collecting the data, connecting with the contributors or procuring relevant information. The team has been an ardent support to the editorial, designing and production team. Their endless efforts to recruit the best for this project, has resulted in the accomplishment of this book. They are a veteran in the field of academics and their pool of knowledge is as vast as their experience in printing. Their expertise and guidance has proved useful at every step. Their uncompromising quality standards have made this book an exceptional effort. Their encouragement from time to time has been an inspiration for everyone.

The publisher and the editorial board hope that this book will prove to be a valuable piece of knowledge for researchers, students, practitioners and scholars across the globe.

List of Contributors

Dalia M.N. Elsheakh, Hala A. Elsadek and Esmat A. Abdallah
Electronics Research Institute, Giza, Egypt

Amir Jafargholi and Manouchehr Kamyab
K. N. Toosi University of Technology, Iran

J. Zhang, S.W. Cheung and T.I. Yuk
Department of Electrical and Electronic Engineering, The University of Hong Kong, Hong Kong, China

Qi-Ye Wen, Huai-Wu Zhang, Qing-Hui Yang, Zhi Chen, Bi-Hui Zhao, Yang Long and Yu-Lan Jing
State Key Laboratory of Electronic Films and Integrated Devices, University of Electronic Science and Technology of China, Chengdu, China

Shah Nawaz Burokur, Abdelwaheb Ourir, André de Lustrac and Riad Yahiaoui
Institut d'Electronique Fondamentale, Univ. Paris-Sud, CNRS UMR 8622, France

Merih Palandöken
Berlin Institute of Technology, Germany

Shiyang Liu and Huajin Chen
Institute of Information Optics, Zhejiang Normal University, China

Zhifang Lin
State Key Laboratory of Surface Physics (SKLSP) and Department of Physics, Fudan University, China

S. T. Chui
Department of Physics and Astronomy, University of Delaware, USA

Amir Jafargholi and Mehdi Veysi
K.N. Toosi University of Technology, Iran

Mahmood Rafaei Booket
Tarbiat Modares University, Iran

Alejandro L. Borja, Angel Belenguer and Joaquin Cascon
Departamento de Ingeniería Eléctrica, Electrónica, Automática y Comunicaciones, Escuela, Politécnica de Cuenca, Universidad de Castilla-La Mancha, Spain

James R. Kelly
School of Electronic, Electrical and Computer Engineering, University of Birmingham, UK

Vicente E. Boria
Instituto de Telecomunicaciones y Aplicaciones Multimedia, Universidad Politécnica de Valencia, Spain